Physics with Answers contains 500 problems covering the full range of introductory physics and its applications to many other subjects, along with clear, step-by-step solutions to each problem. No calculus is required. Students often have difficulty in solving practical problems after a subject is introduced in class. This book bridges the gap – it contains every type of problem likely to be encountered at this level, so by attempting these exercises and learning from the solutions, students will gain confidence in solving class problems and improve their grasp of physics.

The book is split into two parts. The first contains the problems, together with useful summaries of the main results needed for solving them. The second part gives full solutions to each problem, often accompanied by thoughtful comments. Subjects covered include statics, Newton's laws, circular motion, gravitation, electricity and magnetism, electric circuits, liquids and gases, heat and thermodynamics, light and waves, atomic physics, and relativity.

The problems are drawn from many fields, including physics, chemistry, biology, engineering, medicine, and architecture. The book will be invaluable to anyone taking an introductory course in physics, whether at college or pre-university level.

PHYSICS WITH ANSWERS

P HYSICS
W ITH
A NSWERS

500 PROBLEMS AND SOLUTIONS

A. R. King

University of Leicester

O. Regev

Technion – Israel Institute of Technology

CAMBRIDGE
UNIVERSITY PRESS

PUBLISHED BY THE PRESS SYNDICATE OF THE UNIVERSITY OF CAMBRIDGE
The Pitt Building, Trumpington Street, Cambridge CB2 1RP, United Kingdom

CAMBRIDGE UNIVERSITY PRESS
The Edinburgh Building, Cambridge CB2 2RU, United Kingdom
40 West 20th Street, New York, NY 10011-4211, USA
10 Stamford Road, Oakleigh, Melbourne 3166, Australia

First published in 1997

Printed in the United States of America

Typeset in Times Roman

*A catalog record for this book is available from
the British Library*

Library of Congress Cataloging-in-Publication Data

King. A. R.
Physics with answers: 500 problems and solutions/A. R. King.
O. Regev.
p. cm.
ISBN 0-521-48270-4 (hc).—ISBN 0-521-48369-7 (pbk.)
1. Physics—Problems, exercise, etc.   I. Regev, O. II. Title.
QC32.K62   1997   96-2866
530'.076—dc20   CIP

ISBN 0 521 48270 4 hardback
ISBN 0 521 48369 7 paperback

CONTENTS

PREFACE

Physics is the most fundamental of the sciences, and some knowledge of it is required in fields as disparate as chemistry, biology, engineering, medicine, and architecture. Our experience in teaching physics to a wide variety of audiences in the U.S. and Europe over many years is that, while students may acquire some familiarity with formal concepts of physics, they are all too often uneasy about applying these concepts in a variety of practical situations. As an elementary example, they may be able to quote the law of conservation of angular momentum in the absence of external torques, but be quite unable to explain why a spinning top does not fall over. The physicist Richard Feynman coined the phrase "fragile knowledge" to describe this kind of mismatch between knowledge of an idea and the ability to apply it.

In our view there is really only one way of acquiring a robust ability to use physics: the repeated employment of physical concepts in a wide variety of applications. Only then can students appreciate the strength of these ideas and feel confident in using them. This book aims to meet this need by providing a large number of problems for individual study. We think it very important to provide a full solution for each one, so that students can check their progress or discover where they have gone wrong. We hope that users of this book will be able to acquire a working knowledge of those parts of physics they need for their science.

Calculation is an essential ingredient of physics: the ability to make quantitative statements which can be checked by observation and experiment is the basis of the enormous success of modern science and technology. Nevertheless, in this book we have tried to avoid mathematical complications which are not fundamental to understanding the physics. In particular we make no use of calculus. It is worth pointing out that many practical situations that scientists encounter are too complex to allow detailed calculations.

In these cases a simple estimate is often quite sufficient to give great insight, and is in any case an indispensable preliminary to any attempt at a more elaborate treatment.

The book contains problems organized in three chapters, on mechanics, electromagnetic theory, and the properties of matter and waves. We give brief summaries of the relevant theory at the beginning of each of the chapters. These are not extensive, as this is not intended as a textbook, but they do cover all of the topics, and establish the conventions we use. Solutions to each problem are given in the second half of the book. We hope that users of the book will attempt a problem before looking up the solution; even an unsuccessful attempt brings the subject into much sharper focus than simply reading the solution before appreciating the difficulty. Knowledge hard-won in this way is the essence of a working grasp of physics, just as an athlete's performance owes much to long hours of training. Realistically, however, we expect that some of the time this will not happen, particularly when the subject is new. We hope we have provided enough problems so that the reader may, if desired, use the first one or two solutions on any topic to "spot the pattern," and thus acquire the ability to attempt the later problems without having to look up the solution first. Accordingly, there is a general tendency for the problems in a given area to be easier at the beginning than the end. However, we have resisted any idea of doing this absolutely systematically, for the good reasons that (a) the degree of difficulty of a problem is often a rather subjective judgement, and (b) we do not want readers to *expect* the problems to get too difficult for them as the section proceeds. Indeed, we have deliberately sprinkled some simpler problems over the sections to avoid this, so our advice to the reader is always at least to try the problem before giving up!

We hope that this book will be useful to college and university undergraduates in the physical and life sciences, engineering, medicine and architecture, as well as for some high school and secondary school courses. With this in mind we have tried to include problems drawn directly from these subjects. The enormous range of applicability of physics, from understanding why black holes are black to why boiling frankfurters split lengthways, is for us one of its great fascinations, and we hope we have managed to convey a little of this in the book. We hope too that it will provide its readers with the basis of a sound and adaptable knowledge of physics. As a very important side-effect, we trust that it will be useful in preparing for examinations: most common types of physics problems set at this level will be encountered here. We make no apology to our colleagues in universities and schools for this – after all, in an important sense the subject is defined by the huge range of questions it can answer. A student who has acquired the ability to solve problems (and so pass examinations) has a good grounding in physics, and thoroughly deserves success.

NOTE ON UNITS

This books uses SI (meter–kilogram–second) units throughout, with one exception: we follow the customary usage of *gram* moles, rather than kilo-moles, in discussing gases. We sometimes state problems using conventional non-SI units (e.g. km/h for speeds), but these are converted into SI units in the solutions. Numerical answers are usually given to two significant figures.

PHYSICAL CONSTANTS USED IN THIS BOOK

Gravitational constant	$G = 6.7 \times 10^{-11} \ \mathrm{N\,m^2 \ kg^{-2}}$
Acceleration due to gravity	$g = 9.8 \ \mathrm{m\,s^{-2}}$
Speed of light in vacuum	$c = 3 \times 10^8 \ \mathrm{m\,s^{-1}}$
Coulomb constant	$\dfrac{1}{4\pi\epsilon_0} = 9 \times 10^5 \ \mathrm{N \ C^{-2} \ m^{-2}}$
Permeability of vacuum	$\epsilon_0 = 8.84 \times 10^{-12} \ \mathrm{N^{-1} \ C^2 \ m^{-2}}$
Permittivity of vacuum	$\mu_0 = 4\pi \times 10^{-7} \ \mathrm{T\,m\,A^{-1}}$
Boltzmann constant	$k = 1.38 \times 10^{-23} \ \mathrm{J \ K^{-1}}$
Gas constant	$R = 8.31 \ \mathrm{J \ mole^{-1} \ K^{-1}}$ $= 0.082 \ \mathrm{liter \ Atm \ K^{-1}}$ $= 8.31 \times 10^3 \ \mathrm{J\,kg^{-1} \ K^{-1}}$
Specific heat of water	$C_w = 4200 \ \mathrm{J\,kg^{-1}\,{}^\circ C^{-1}}$
Planck constant	$h = 2\pi\hbar = 6.63 \times 10^{-34} \ \mathrm{J\,s}$
Proton charge	$e = 1.6 \times 10^{-19} \mathrm{C}$
Mass of electron	$m_e = 9.1 \times 10^{-31} \ \mathrm{kg}$
Mass of proton	$m_p = 1.67 \times 10^{-27} \ \mathrm{kg}$
Atomic mass unit	$m_H = 1.67 \times 10^{-27} \ \mathrm{kg}$
Compton wavelength	$\lambda_c = 2.4 \times 10^{-12} \ \mathrm{m} = 0.024 \mathrm{\mathring{A}}$
Rydberg	$E_0 = 13.6 \ \mathrm{eV}$

PART ONE

PROBLEMS

CHAPTER ONE

MECHANICS

■ SUMMARY OF THEORY

I. Status of the Subject

Newtonian mechanics provides a complete description of virtually all mechanical phenomena. The two exceptions to this statement concern (a) speeds approaching that of light, and (b) lengths of order the size of atoms.

Note that air resistance is neglected in all problems unless the contrary is explicitly stated.

2. Statics

● Equilibrium of a body under external forces requires that their resultant is zero, i.e.

$$\Sigma F_x = \Sigma F_y = \Sigma F_z = 0, \tag{1}$$

where F_x, F_y, F_z are the three Cartesian components of the resultant force. If the forces act on lines that all meet at a point, this condition is also sufficient. It is then legitimate to represent all the forces as acting at the body's center of mass.

● The *center of mass* is the pointd with coordinates (x_{CM}, y_{CM}, z_{CM}), where

$$x_{CM} = \frac{\Sigma m_i x_i}{\Sigma m_i}, \tag{2}$$

etc. Here the summations extend over all the mass points of the body. The position of the center of mass can often be found from symmetry require-

3

ments. If two bodies of mass m_1, m_2 are joined together so that their centers of mass have coordinates $(x_1, y_1, z_1), (x_2, y_2, z_2)$, the center of mass of the combined body has coordinates given by applying (2) to them, i.e.

$$x_{CM} = \frac{m_1 x_1 + m_2 x_2}{m_1 + m_2}, \text{ etc.} \tag{3}$$

● If the external forces do not act along lines meeting at a point, we require in addition to (1) that the resultant torque should vanish. In this book we restrict attention to forces acting in a plane, and the torque condition for equilibrium is

$$\Sigma M_O = 0, \tag{4}$$

where M_O is the product of the force and its perpendicular distance from the axis through O. The torque is counted positive if the force tends to cause anticlockwise rotation about the axis and negative otherwise. The position O of the axis may be chosen freely: if there is an unknown force in the problem, it is generally useful to choose O on the line of action of this force, so that its torque vanishes. Given a point O such that $\Sigma M_O = 0$, then $\Sigma M_{O'} = 0$ for any other point O'.

● The *frictional force f* or F_f acting on a body has two forms: if the body is static, and the normal reaction force between two surfaces is N, then f takes a value no larger than a certain maximum, i.e.

$$f \leq \mu_s N. \tag{5}$$

Here μ_s is a dimensionless quantity characteristic of the two surfaces, called the coefficient of static friction. Note that this equation does not *determine* the actual value of f: this is found from the equilibrium conditions (1, 4). If the force required to maintain equilibrium exceeds $\mu_s N$, the bodies slide with respect to each other, and the frictional force becomes

$$f = \mu N, \tag{6}$$

where μ is now the coefficient of sliding (or kinetic) friction.

3. Kinematics

● *Average speed* = (distance traveled)/(time).

● In adding two *velocities* (u_x, u_y, u_z) and (v_x, v_y, v_z), we must add component by component, i.e. the resultant velocity is $(u_x + v_x, u_y + v_y, u_z + v_z)$. This form of addition (and subtraction) also applies to accelerations, momenta, etc. and expresses what is sometimes called the *parallelogram* (or *triangle*) rule (see the Figure for the case of adding two vectors **A**, **B** in the plane).

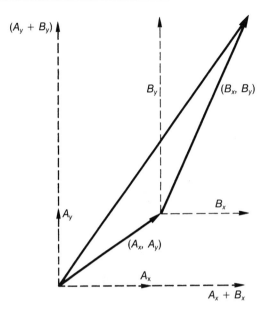

● The *relative velocity* of a moving point A with respect to a moving point B, whose velocities in a given reference frame are $(u(A)_x, \ldots, \ldots)$, $(u(B)_x, \ldots, \ldots)$ is given by subtracting B's velocity from A's component by component, i.e. by $([u(A)_x - u(B)_x], \ldots, \ldots)$.

● *Acceleration* = (change of velocity)/(time).

Note that zero acceleration does *not* automatically imply zero velocity: steady motion has zero acceleration.

● Under *constant acceleration a*, the velocity v and distance x traveled are related to the elapsed time t and initial velocity v_0 by the three formulae

$$v = v_0 + at, \tag{7}$$

$$v^2 = v_0^2 + 2ax, \tag{8}$$

$$x = v_0 t + \frac{at^2}{2}. \tag{9}$$

In two- or three-dimensional motion these formulae can be used component by component. If air resistance is neglected, projectiles have constant vertical acceleration and zero horizontal acceleration.

4. Newton's Second Law

● The fundamental postulate of Newtonian mechanics explains what happens when the resultant external force on a body does not vanish as in statics: *the resultant force on a body equals the rate of change of its momentum.* Here momentum = mass × velocity. If the mass of the body does not change (true for all the problems in this book), we can write Newton's second law in the familiar form

$$\Sigma F_x = ma_x, \tag{10}$$

$$\Sigma F_y = ma_y, \tag{11}$$

$$\Sigma F_z = ma_z. \tag{12}$$

These equations give us the accelerations in terms of the forces. Kinematics can then be used to find the motion.

5. Work, Energy, and Power

● *Work* = (force) × (distance moved in direction of force)

Thus if the motion makes angle θ to the force F, the work done by the force in moving distance l is

$$W = Fl \cos \theta. \tag{13}$$

(Here it is assumed that the force F does not change during the motion through l.)

● *Power* = rate of working. Thus, if work W is performed at a uniform rate in time t, the power is

$$P = \frac{W}{t}. \tag{14}$$

● A body of mass m moving with velocity v has *kinetic energy*

$$T = \frac{1}{2}mv^2. \tag{15}$$

● If the body is raised through a height h against the Earth's gravity, it gains *gravitational potential energy*

$$U = mgh. \tag{16}$$

● The principle of *conservation of energy* states that the total energy of a closed system remains constant. If the only forces acting on a mechanical system are *conservative*, no mechanical energy is converted to other forms, and the total *mechanical* energy is conserved. The commonest example of a conservative force is gravity: a body moving under gravity alone conserves the sum of its kinetic and potential energies, i.e.

$$T + U = \tfrac{1}{2}mv^2 + mgh = \text{constant}. \tag{17}$$

Forces which are not conservative (e.g. friction) and convert mechanical energy to heat are called *dissipative*.

6. Impulse and Momentum

● It follows from Newton's second law that *the total momentum of an isolated system remains constant*, i.e.

$$\Sigma mv_x = \text{constant}, \tag{18}$$

$$\Sigma mv_y = \text{constant}, \tag{19}$$

$$\Sigma mv_z = \text{constant}, \tag{20}$$

where the summation is over all the bodies of the system.

In some cases we deal with systems where bodies move freely except for large forces F, which act for short times t (e.g. collisional forces). In these cases it is easier to deal with the product $I = Ft$, which is called the *impulse*. From Newton's second law it follows that the total impulse on a body gives the change of its momentum.

In collision problems, the effects of the elastic forces of collision are expressed in the *coefficient of restitution e*, defined by

(relative velocity after collision) = $-e \times$ (relative velocity before collision).

If $e = 1$, the collision is *elastic* and total mechanical energy is conserved. If $e < 1$, the collision is *inelastic* and some of the mechanical energy is lost in the collision, e.g. as heat, deformation of the bodies, etc.

7. Circular motion

● The *angular velocity* of a point mass about another point is defined as

$$\omega = \frac{v}{r} \tag{21}$$

where v is the linear velocity of the mass perpendicular to the line joining the two points, and r is the length of this line. Clearly, a rigid body rotates with uniform angular velocity about any of its points.

● For a body to move in a circle of radius r with speed v requires *centripetal acceleration*

$$a_c = \frac{v^2}{r} = \omega^2 r \qquad (22)$$

directed towards the center of the circle. By Newton's second law this requires a *centripetal force*

$$F_c = \frac{mv^2}{r} = m\omega^2 r \qquad (23)$$

directed towards the center of the circle, where m is the mass of the body.

● *Angular acceleration* α = rate of change of angular velocity. If the angular velocity changes by ω at a uniform rate in time t, we have

$$\alpha = \frac{\omega}{t}. \qquad (24)$$

If α is constant, there is a complete analogy with the case of constant linear acceleration a, and the three formulae given for that case can be taken over with the substitution of α for a, ω for v and the angular displacement θ for x.

● Newton's second law applied to rotational motion of a particle of mass m about a fixed point O implies that

$$\Sigma M_O = mr^2 \alpha. \qquad (25)$$

Thus if the total torque about O vanishes, the *angular momentum $mr^2\omega$* is conserved.

8. Harmonic motion

A body is undergoing *simple harmonic motion* when it moves in a straight line under a restoring force proportional to the distance x from a fixed point. The acceleration of such a body can be expressed as

$$a = -\omega^2 x. \qquad (26)$$

Here ω is the *angular frequency*. The concept can be extended to angular motion. The motion repeats itself exactly after a time

$$P = \frac{2\pi}{\omega}. \qquad (27)$$

P is called the *period*. The maximum displacement from the center of force (e.g. $x = 0$) is called the *amplitude*. The period of a *simple pendulum*, a mass suspended from a string of length l oscillating under gravity, is

$$P = 2\pi \left(\frac{l}{g}\right)^{1/2}, \tag{28}$$

independent of the mass and the amplitude of the motion, provided that this remains small. The period of a mass m moving on a smooth horizontal table attached to a spring of constant k whose other end is fixed is

$$P = 2\pi \left(\frac{k}{m}\right)^{1/2}. \tag{29}$$

● If simple harmonic motion of angular frequency ω is initiated from rest with displacement x_0, the subsequent displacement is

$$x(t) = x_0 \cos \omega t. \tag{30}$$

If simple harmonic motion of angular frequency ω is initiated from the origin with speed v_0, the subsequent displacement is

$$x(t) = \frac{v_0}{\omega} \sin \omega t. \tag{31}$$

9. Gravitation

● Newton's law of *universal gravitation* states that the attractive gravitational force between two point masses m_1, m_2 is

$$F_{\text{grav}} = \frac{Gm_1 m_2}{d^2} \tag{32}$$

where G is a universal constant, and d is the separation of the two masses. The gravitational potential energy of the two masses is

$$U = -\frac{Gm_1 m_2}{d}. \tag{33}$$

It can be shown that the gravitational force exerted by a uniform sphere is the same as if the sphere's mass were all concentrated at its center.

For bodies close to the Earth, d is always effectively equal to the Earth's radius R_e, so the downwards vertical force on a body of mass m is

$$F_{\text{grav}} = mg, \tag{34}$$

where $g = GM_e/R_e^2$, with M_e = mass of the Earth. Here g is called the *surface gravity* or the *acceleration due to gravity*. If the body is subject to upwards vertical acceleration a, we define the *effective gravity* as

$$g_{\text{eff}} = g + a. \tag{35}$$

For example, at the Earth's equator, some of the gravitational force must be used to provide the centripetal acceleration needed to keep the body on the Earth's surface, so the effective gravity is lower there.

10. Motion of a rigid body

● A rigid body is one in which the distances between any of its particles remain constant at all times.

● The motion of a rigid body can be decomposed into the linear motion of its center of mass, and rotations about the center of mass. The center of mass motion is that of a point object of the same mass as the body. As explained above, a rigid body has uniform angular velocity about any point.

● If Newton's second law is applied to rotational motion about *either* any fixed point O *or* the center of mass, it implies that

$$\Sigma M_O = I\alpha, \tag{36}$$

where

$$I = \Sigma mr^2, \tag{37}$$

is called the *moment of inertia* about O. Here r is the perpendicular distance of each point of mass m from the axis. The moment of inertia plays for angular motion the role of the mass in linear motion. The moments of inertia of simple bodies may be found easily, and are given in Table 1.

● If the total torque about O vanishes, then the *angular momentum $I\omega$* is conserved. This is the analog of the conservation of (linear) momentum for an isolated system referred to in Section 6 above.

The kinetic energy of rotation with angular velocity ω about a point O is

$$T = \frac{1}{2}I\omega^2, \tag{38}$$

where I is the relevant moment of inertia. The rate of increase of T is given by the *work done by the torques M_O*, which is $\Sigma M_O\theta$, where θ is the angle traveled in the direction of the torque.

The period of a *physical pendulum* undergoing simple harmonic motion is

$$P = 2\pi \left(\frac{I}{mgl_{\text{CM}}}\right)^{1/2}, \tag{39}$$

where I is the relevant moment of inertia, m the mass of the body, and l_{CM} is the distance of the center of mass from the pivot.

TABLE 1. Moments of inertia of simple uniform bodies of mass M about their symmetry axes.

body	I
circular hoop, radius r cylindrical shell, radius r	Mr^2
circular disc, radius r solid cylinder, radius r	$\frac{1}{2} Mr^2$
rod, length l	$\frac{1}{12} Ml^2$
sphere, radius r	$\frac{2}{5} Mr^2$

■ STATICS

P1. Show that the center of mass of the Earth–Sun system is located *inside* the Sun. (The Sun's mass $M_\odot = 2 \times 10^{30}$ kg, the Earth's mass $M_e = 6 \times 10^{24}$ kg, the Sun's radius $R_\odot = 7 \times 10^8$ m, and the Earth–Sun distance $d_e = 1.5 \times 10^{11}$ m.) Where is the center of mass of the Sun–Jupiter system? (Jupiter's mass $M_J = 2 \times 10^{27}$ kg, Jupiter–Sun distance $d_J = 1.4 \times 10^{12}$ m.)

P2. A tennis racket can be approximated by a circular hoop of radius r and mass m_1 attached to a uniform shaft of length l and mass m_2. Assuming that $r = l/2$ and $m_1 = m_2 = m$, find the position of the racket's center of mass.

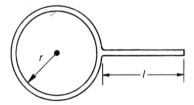

P3. The tennis racket of the previous question is modified by adding a point mass $m_3 = m/2$ to the part of the rim furthest from the shaft. Find the new position of the center of mass.

P4. A pizza can be regarded as a uniform thin disk of radius r and mass m. A narrow slice of angle $\theta = 20°$ is cut out and eaten. Approximating the slice as a triangle, where would you have to support the partly eaten pizza to hold it in balance?

P5. Ships which have been emptied of cargo are often refilled with ballast (e.g. sand, water). Why?

P6. Given two sets of weighing scales and a long board, how could you determine the position of the center of mass of a person?

P7. A mass rests on an inclined plane of angle $\theta = 30°$. The coefficient of static friction is $\mu_s = 0.6$. Draw a diagram showing all the forces acting on the mass, and explain their origin. Calculate their values if the mass is $m = 5$ kg. Verify that under these conditions the mass will not slide.

P8. A mass $m = 10$ kg hangs by two strings making angles $\alpha = 45°$ and $\beta = 60°$ to the vertical. The strings are connected through pulleys to two masses m_1 and m_2 (see Figure). Find m_1, m_2 such that the mass hangs in equilibrium.

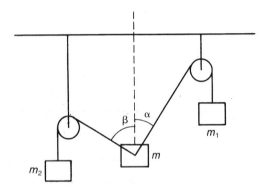

P9. A uniform sphere of mass m and radius r hangs from a string against a smooth vertical wall, the line of the string passing through the ball's center (see Figure). The string is attached at a height $h = \sqrt{3}r$ above the point where the ball touches the wall. What is the tension T in the string, and the force F exerted by the ball on the wall? If the wall is rough, with coefficient of static friction μ_s, are these forces increased or reduced?

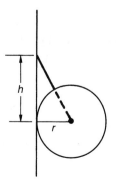

P10. A circus performer of mass $m = 60$ kg stands at the midpoint of a rope of unstretched length $l_0 = 6$ m. It is known that the tension T in the rope is proportional to the amount it is stretched, i.e. $T = \kappa(l - l_0)$, where κ is a constant and l the actual length of the rope. How large must κ be if the performer is not to sink more than a distance $h = 1$ m below the endpoints of the rope? With this value of κ, how much would the rope extend if the performer were to release one end of it and hang vertically from it?

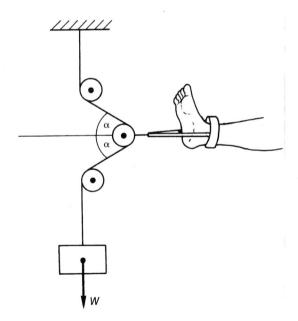

P11. A mass m is suspended from the center of a wire, which is stretched over two supports of equal heights. The tensions at each end of the wire are T. Show that however large T is made, the wire is never completely horizontal. Estimate the angle to the horizontal if $T = 100mg$.

P12. A patient's leg is in traction with the arrangement shown in the Figure, with $W = 100$ N. A student nurse moves the cord to an anchoring point nearer to the patient, so that the two angles of the cord to the horizontal change from $\alpha_1 = 45°$ to $\alpha_2 = 30°$. Does this make any difference to the patient?

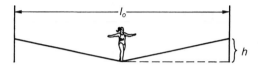

P13. The human forearm can be approximated by a lever as shown in the Figure. Given that $L = 20l$ and the arm weight is w, what muscle force F must be exerted to lift a weight W with the arm at angle θ to the horizontal? Why is it larger than $W + w$?

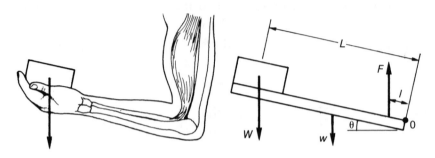

P14. A box of mass m is pulled by a man holding a rope at an angle θ to the horizontal. A second man pulls horizontally in the opposite direction with a force equal to twice the box's weight. What is the maximum value θ_c of θ such that the box begins to move in the direction of the first man without being lifted from the ground? What, in terms of mg, is the force P then exerted by the first man?

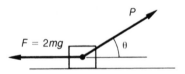

P15. A uniform rod of mass m can rotate freely around a horizontal axis O at one end which is fixed to the floor. It is supported at an angle $\alpha = 45°$ to the floor by a string attached to the other end making an angle $\beta = 15°$ to the vertical,

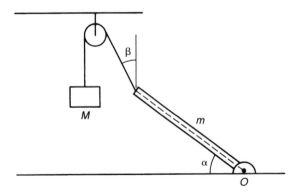

the free end of the string hanging vertically from a pulley and holding a mass M (see Figure). Find M in terms of m if the system is in equilibrium. Calculate the force P exerted by the floor on the axis O, and its direction. (Express your answer in terms of m and g.)

P16. A shop puts up a signboard of mass m hanging from the end of a rod of length l and negligible mass, which is hinged to the shop wall at an axis O. The rod is held horizontal by means of a wire attached to its midpoint and to the wall, a height h above the hinge (see Figure). If the wire will break when its tension T reaches $T_{max} = 3mg$, what is the minimum height h_{min} (in terms of l) that the wire must be attached to the wall?

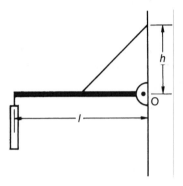

P17. A rectangular door of mass M, width w and height $h = 3w$ is supported on two hinges located a distance $d = w/4$ from its upper and lower edges. If the hinges are arranged so that the upper one carries the entire weight of the door, find the forces (in terms of Mg) exerted on the door by the two hinges.

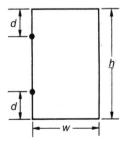

P18. A uniform rod of mass m leans against a smooth vertical wall, making an angle θ_1 with it. Its other end is supported by a smooth plane inclined at an angle θ_2 to the horizontal (see Figure). Find a relation between the angles

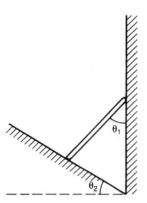

θ_1, θ_2. If $\theta_2 = 30°$ find the forces exerted by the wall and inclined plane on the rod in terms of *mg*.

P19. A uniform ladder leans against a smooth vertical wall, making an angle θ with a horizontal floor. The coefficient of static friction between the ladder and the floor is μ. Find (in terms of μ) the minimum angle θ_m for which the ladder does not slip.

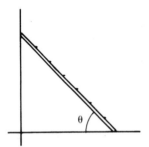

P20. In the configuration of the previous problem, a repair worker whose mass is twice that of the ladder wishes to climb to its top. What does the minimum angle θ_m become?

P21. A uniform rectangular platform of width L hangs by two ropes making angles $\theta_1 = 30°, \theta_2 = 60°$ to the vertical. A load of twice the mass of the platform is placed on it to keep it horizontal: how far from the edge of the platform must it be?

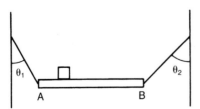

P22. A uniform circular cylinder of radius r has its base on a plane inclined at angle θ to the horizontal. The coefficient of static friction is μ_s. Find the minimum height h of the cylinder such that it overturns rather than sliding.

P23. The human jaw is worked by two pairs of muscles, positioned on each side of the pivot (see Figure). Is it possible to arrange for there to be no reaction force on the pivot when the jaw exerts a steady chewing force C upwards and the lower muscle pair exerts a force L as shown? Find C in this case if the upper and lower muscle pairs act at angles $\theta_u = 50°, \theta_l = 40°$ to the horizontal.

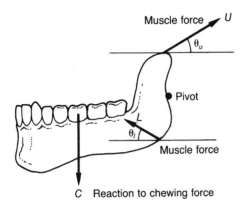

Muscle force U

θ_u

Pivot

L

θ_l

Muscle force

C Reaction to chewing force

P24. A horizontal force $F = 0.2$ N acts on the tooth shown in the Figure. Find the forces F_1, F_2 exerted by the jawbone on the root and vice versa, if $l_1 = 1.5$ cm, $l_2 = 2$ cm.

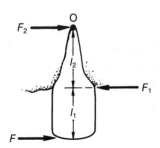

O

F_2

l_2

F_1

l_1

F

P25. A football player of height h is subjected to a horizontal push at his shoulders, which are a distance $h/4$ from his center of mass, which in turn is a distance $5h/8$ from his feet (see Figure). To counteract the push he leans forward at an angle θ to the vertical. The coefficient of static friction between the player's feet and the pitch is μ. Find the minimum angle θ_m of lean such

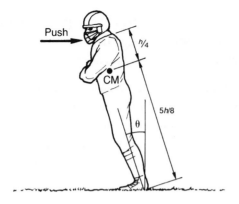

that the player slides backwards rather than being overturned by a strong push.

P26. A woman of mass m stands on one platform of a large beam balance and pulls on a cord connected to the center of its nearer arm. The other platform holds a mass M. What restrictions on M, m are required if the balance is to remain level?

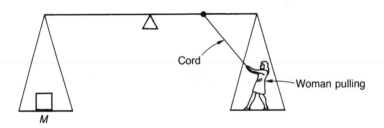

P27. A woman lifts a mass M by means of the double pulley arrangement shown in the Figure. If all sections of the rope are regarded as vertical, the pulleys are very light and friction is negligible, what force must she exert? If she wishes to raise the mass through a height h, what length of rope must she pull down?

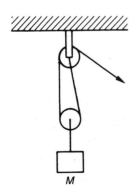

P28. What happens to the results of the previous question if a second pair of pulleys are added, as shown in the Figure?

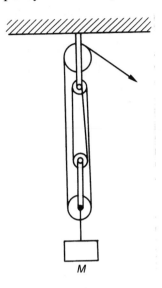

P29. Rotation of the shaft of the right-hand lever (of length b) in the Figure is resisted by a frictional torque whose maximum possible value is G_1. What torque must be supplied to the shaft of the left-hand lever (length a) in order to begin to turn it anticlockwise as shown? Repeat the calculation if the levers are replaced by steadily turning gear wheels as shown. If the left-hand shaft is rotated with angular velocity Ω, what is the angular velocity of the right-hand shaft?

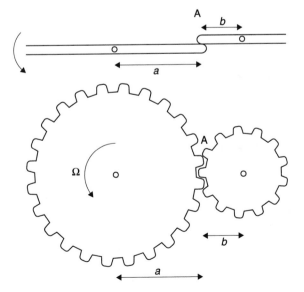

P30. A cylindrical oil drum of mass m and radius R lies on a road against the curb, which has height $R/2$ (see Figure). It is to be lifted gently (quasistatically) on to the sidewalk by means of a rope wound around its circumference. What is the minimum force F_m needed if the rope is pulled horizontally? What is the magnitude and direction of the reaction force at the curb? Will the minimum force F_m change as the drum is lifted? If the rope is pulled at an angle θ to the horizontal, for what value of θ is the required force a minimum? What is the value of this minimum force?

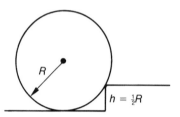

P31. A drinking straw of length l is placed in a smooth hemispherical glass of radius R resting on a horizontal table. Find its equilibrium position
   (a) – if $l < 2R$,
   (b) – if $l > 2R$, assuming that the straw does not fall out.

P32. A woman lifts a mass M slowly by means of a pulley, placed at the height of her hand (see Figure). Her forearm is $f = 24$ cm long, and her biceps muscles are attached to it $a = 3$ cm from the elbow joint. Estimate the tension T in her biceps if her upper arm and forearm make angles θ, ϕ to the vertical. If she keeps $\theta = \phi$, does it get easier or harder to lift the mass as she raises it?

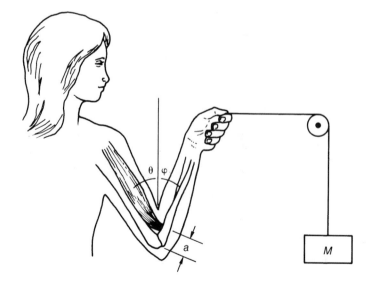

P33. A market stallholder erects an awning (see Figure) of mass M and breadth $2l$. The supports are placed a distance a from the rear edge, which is secured by a vertical rope. Find the force F on the supports. If instead a second set of supports is placed a distance a from the front of the awning and the rope is removed, what is the new force F on each of the sets of supports? Compare the two cases if $M = 50$ kg, $l = 1$ m, $a = 10$ cm.

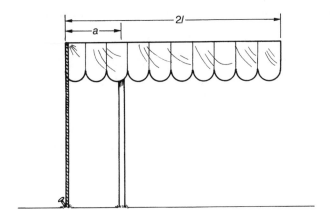

KINEMATICS

P34. A train travels 50 km in half an hour. It then stops at a station for 20 minutes, before traveling for 2 hours at an average speed of 90 km/h. What was the train's average speed over the whole journey?

P35. A car starts from rest and reaches a velocity of 100 km/h after accelerating uniformly for 10 s. What distance has it traveled? What was its average velocity?

P36. A train travels a distance s in a straight line. For the first half of the distance its velocity has the constant value v_1, and for the second half it has the constant value v_2. What is the average velocity? Is it larger or smaller than $(v_1 + v_2)/2$?

P37. A police officer on a motorcycle chases a speeding car on a straight highway. The car's speed is constant at $v_c = 120$ km/h, and the officer is a distance $d = 500$ m behind it when she starts the chase with velocity $v_p = 180$ km/h. What is the police officer's speed relative to the car? How long will it take her to catch up with it?

P38. Taking off from a point on the Equator in the late afternoon and flying due West, passengers on the Concorde supersonic airliner see the sun set and then rise again ahead of them. Estimate Concorde's minimum speed. (Earth's radius = 6400 km.)

P39. The maximum straight-line deceleration of a racing car under braking is $5\,\mathrm{m\,s^{-2}}$. What is the minimum stopping distance of the car from a velocity of 100 km/h? What does this distance become if the velocity has twice this value?

P40. A rocket-powered sled accelerates from rest. After $t = 10$ s it has traveled a distance $x = 400$ m. What is its speed in km/h at this point?

P41. A ball is thrown vertically upwards with initial speed 10 m s^{-1} from the edge of a roof of height $H = 20$ m. How long does it take for the ball to hit the ground? At what velocity does it hit the ground?

P42. A stone is dropped from rest into a well. It is observed to hit the water after 2 s. Find the distance down to the water surface. How fast must the stone be thrown downwards in order to hit the surface after only 1 s? What are the impact velocities in the two cases?

P43. A car and a truck start moving at the same time, but the truck starts some distance ahead. The car and the truck move with constant accelerations $a_1 = 2\ \mathrm{m\,s^{-2}}$, $a_2 = 1\ \mathrm{m\,s^{-2}}$ respectively. The car overtakes the truck after the latter has moved 32 m. How long did it take the car to catch up with the truck? What were the velocities of the car and the truck at that moment? How far apart did the truck and the car start?

P44. A rocket climbs vertically and is powered in such a way that it has constant acceleration a. It reaches a height of 1 km with a velocity of 100 m s^{-1}. What is the value of a? How long does the rocket take to reach this 1 km height?

P45. A bullet is fired vertically from a toy pistol with muzzle velocity 30 m s^{-1}. How high above the firing point does the bullet go before falling back under gravity? What is its velocity 4 s after being fired? At what height is it then?

P46. A body falls freely from rest to the ground a distance h below. In the last 1 s of its flight it falls a distance $h/2$. What is h?

P47. A man falls from rest from the top of a building of height $H = 100$ m. A time $t = 1$ s later, Superwoman swoops after him with initial speed v_0 downwards, subsequently falling freely. She catches the man at a height $h = 20$ m above the ground. What was v_0?

P48. A boy in an elevator throws a ball vertically upwards with speed $v_0 = 5$ m s^{-1} relative to the elevator. The elevator has constant upward acceleration $a = 2$ m s^{-2}. How long does it take for the ball to return to the boy's hand?

P49. An artillery shell is fired from a cannon with an elevation of $\alpha = 30°$ and muzzle velocity of $v_0 = 300$ m s^{-1}. Find the time of flight of the shell, and its range.

P50. A certain athlete consistently throws a javelin at a speed of 25 m s^{-1}. What is her best distance? On one occasion the athlete released the javelin poorly, and achieved only one half of this distance. At what elevation angle did she release the javelin?

P51. In the last problem, does the elevation angle for half distance depend on the speed of the throw? Explain your answer.

P52. A projectile is fired on level ground. Show that, for given range and initial velocity the projection angle has two possible values, which are symmetrically spaced each side of 45°.

P53. In the movie *Speed* a bus has to leap a gap in an elevated freeway. If the bus had speed $v_0 = 100$ km/h and the gap was $x = 15$ m,
  (a) – assuming the takeoff and landing points were at the same level, find the angle of projection of the bus's center of mass;
  (b) – if the bus took off horizontally, how much lower must the landing side have been than takeoff?

P54. A rifleman aims directly and horizontally at a target at distance x on level ground, and his bullet strikes a height h too low. If $h \ll x$, show that in order to hit the target, he should aim a height h *above* it.

P55. A transport airplane flies horizontally with a constant velocity of 600 km/h, at a height of 2 km. Directly over a marker it releases an empty fuel tank. How far ahead of the marker does the tank hit the ground? At this time, is the airplane ahead or behind the tank?

P56. An airplane in steady level flight with velocity $v = 700$ km/h releases a number of bombs at regular intervals $\Delta t = 1$ s. A photograph of the release is taken from an accompanying airplane. Describe the relative position of the first airplane and the bombs on the photograph. How far apart are the impact points of the bombs on the ground?

P57. A combat tank fires a shell while moving on horizontal ground with velocity $u = 10$ m s^{-1}. The gun is pointing directly forwards with elevation $\alpha = 5°$, and the muzzle velocity is $v_0 = 1000$ m s^{-1}. The shell hits a target which is moving directly away from the tank at $w = 15$ m s^{-1}. How far from the tank is the target at the moment of impact? How far apart were the tank and the target when the shell was fired?

P58. A softball is thrown at an angle of $\alpha = 60°$ above the horizontal. It lands a distance $d = 2$ m from the edge of a flat roof, whose height is $h = 20$ m; the

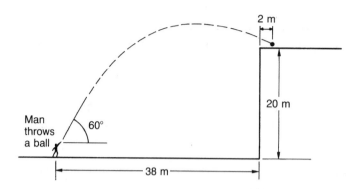

edge of the roof is $l = 38$ m from the thrower (see Figure). At what speed was the softball thrown?

P59. A projectile is launched with horizontal and vertical velocity components u, v. Show that its trajectory is a parabola, and that the maximum height and the range (on level ground) are $h = v^2/g, r = 2uv/g$, respectively.

P60. An athlete can throw the javelin at four times the speed at which she can run. At what angle in her reference frame should she launch the javelin for maximum range?

P61. A small boy uses a pea-shooter to blow a pea directly at a cat in a tree. The cat is startled by the noise of the boy blowing and falls vertically out of the tree. Does the pea miss?

P62. A downhill skier approaches horizontally a hump of height $h = 1$ m which levels out before steepening suddenly to an angle $\alpha = 25°$ to the horizontal (see Figure). If her horizontal speed at the top of the hump is $u = 100$ km/h, how long does she spend in the air before landing down the slope? If the skier is able to jump vertically at speed $v = 5$ m s^{-1}, and she moves more quickly when in contact with the snow than in the air, can you suggest a strategy for improving her time?

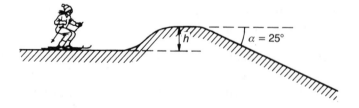

P63. A man can swim at a speed $v_s = 1$ m s^{-1}, and wishes to cross a river of width $L = 100$ m flowing at $v_w = 0.5$ m s^{-1} to reach his girlfriend who is directly opposite him on the other bank. In what direction should the man swim so as to reach her as soon as possible? How long will it take him?

P64. Two trains A and B are traveling in opposite directions along straight parallel tracks at the same speed $v = 60$ km/h. A light airplane crosses above them. A person on train A sees it cross at right angles, while a person on train B sees it cross the track at an angle $\theta = 30°$. At what angle α does the airplane cross the track as seen from the ground? What is its ground speed v_g?

P65. Rain falls vertically at speed u on a man who runs at horizontal speed v. Show that he sees the rain falling towards him at speed $(u^2 + v^2)^{1/2}$ and angle $\phi = \tan^{-1} v/u$ to the vertical. The man leans into the rain as he runs, at angle θ to the vertical. His total frontal area is A_f, and his total area viewed from above is A_t. If $A_t < A_f$, show that he gets least wet if he leans so that $\theta = \phi$. If he runs a distance l and there is mass ρ of water per unit volume of rain, show that he absorbs a minimum total mass

$$m = A_t l \rho \frac{(u^2 + v^2)^{1/2}}{v} \tag{40}$$

of water.

P66. A car rounds a bend in a road at a speed of 70 km/h and collides with a second car that has emerged from a concealed side road 50 m from the bend. Analysis of the damage to the cars shows that the collision took place at a closing speed of 10 km/h or less. In making his insurance claim, the driver of the first car asserts that the second car emerged from the side road in such a way that the first car had only 4 m in which to brake. Is this version plausible?

■ NEWTON'S SECOND LAW

P67. A mass $m_1 = 1$ kg lies on a smooth table and is attached by a string and a frictionless pulley to a mass $m_2 = 0.01$ kg hanging from the edge of the table (see Figure). The system is released from rest. Calculate the distance the mass m_1 moves across the table in the first 10 s. How long will it take for this mass to travel 1 m from its initial position?

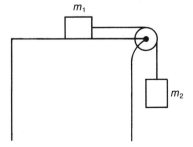

P68. A mass $m = 20$ kg is pulled upwards with constant acceleration by a cable attached to a motor. The cable can withstand a maximal tension of 500 N. What is the maximum acceleration a_{max} possible? If the acceleration has this maximum value, what distance will the mass have moved after 2 s, if it starts from rest?

P69. A smooth inclined plane has a slope of 30°. A body begins to move upwards with initial velocity 5 m s^{-1}. How long does it take for the body to begin to slide down the plane again?

P70. Two bodies are attached to the ends of a string hanging from a frictionless pulley (see Figure). The masses of the two bodies are $m_1 = 5$ kg and $m_2 = 10$ kg. Find the accelerations of the masses and the tension in the string.

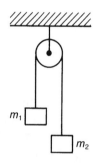

P71. A subway train has constant acceleration $a = 0.1g$. In one of the cars a mass m hangs from the ceiling by means of a string. Find the angle the string makes to the vertical and the tension in the string in terms of m and g.

P72. An elevator of mass M moves upwards with constant acceleration $a = 0.1g$, pulled by a cable. What is the normal force exerted by the elevator floor on a person of mass m standing inside it? What is the tension in the cable? Express your answer in terms of M, m, g.

P73. Two masses m, M lie on each side of a smooth wedge (see Figure), connected by a string passing over a frictionless pulley. The wedge faces make angles $\theta_1 = 53°$ and $\theta_2 = 47°$ to the horizontal respectively. What value must the ratio M/m take so that the masses remain stationary? What is the tension in the string in this case, in terms of m, g?

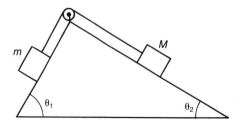

P74. An experiment is performed to determine the value of the gravitational acceleration g on Earth. Two equal masses M hang at rest from the ends of a string on each side of a frictionless pulley (see Figure). A mass $m = 0.01M$ is placed on the left-hand mass. After the heavier side has moved down by $h = 1$ m the small mass m is removed. The system continues to move for the next 1 s, covering a distance of $H = 0.312$ m. Find the value of g from these data.

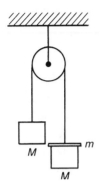

P75. A rifleman holds his rifle at a height $h = 1.5$ m and fires horizontally over level ground. The bullet lands at a distance $s = 500$ m from the muzzle of the gun. What was the muzzle velocity of the bullet? The rifle barrel has length $l = 0.5$ m. Assuming that the bullet has constant acceleration inside it, calculate the force on the bullet, if its mass was 10 g.

P76. A skydiver jumps from an airplane and acquires a falling velocity of $20\ \text{m s}^{-1}$ before opening her parachute. As a result her falling velocity drops to $5\ \text{m s}^{-1}$ in 5 s. The skydiver has mass $m = 50$ kg. Assuming that the deceleration was constant, find the total tension in the parachute cords and the resultant force on the skydiver.

P77. The coefficient of sliding friction between the tires of a car and the road surface is $\mu = 0.5$. The driver brakes sharply and locks the wheels. If the velocity of the car before braking was $v_0 = 60$ km/h, how much time will the car take to stop? What is the stopping distance?

P78. The coefficient of kinetic friction between a sled of mass $m = 10$ kg and the snow is $\mu = 0.1$. What horizontal force F is required to drag the sled at a constant velocity?

P79. A skier is stationary on a ski slope of angle $\alpha = 15°$. The pressure of his skis gradually melts the snow and reduces the effective coefficient of static friction μ_s. What is the value of this coefficient at the moment that the skier begins to move? If the coefficient μ of kinetic friction between the skis and the snow is 0.1, what is his velocity after 5 s, and what distance has he then traveled?

P80. A length of timber of mass $M = 100$ kg is dragged along the ground with a force $F = 300$ N by means of a rope. The rope makes an angle of $\alpha = 30°$ to the ground. The coefficient of friction between the timber and the ground is $\mu = 0.2$. Find the acceleration a of the timber. Find also the normal force N exerted by the ground on the timber.

P81. A body is given an initial sliding velocity $v_0 = 10$ m s^{-1} up an inclined plane of slope $\alpha = 20°$ to the horizontal. The coefficient of friction is $\mu = 0.2$. Find the time t_{up} the body spends sliding up the slope before reversing its motion, the distance s traveled to this point, and the time t_{down} to return to the starting point.

P82. A mass m is placed on a rough inclined plane and attached by a string to a hanging mass M over a frictionless pulley (see Figure). The angle α of the slope is such that $\sin \alpha = 0.6$. The coefficient of static friction between the mass m and the plane is $\mu_s = 0.2$. Show that equilibrium is possible only if M lies between two values M_1, M_2 and find the values of M_1, M_2 in terms of m.

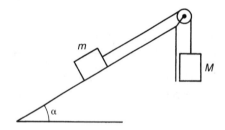

P83. A uniform chain of total length l lies partly on a horizontal table, with a length l_1 overhanging the edge. If μ_s is the coefficient of static friction, how large can l_1 be if the chain is not to slide off the table?

P84. Two equal masses lie on each side of a rough wedge, connected by a string passing over a frictionless pulley. The wedge faces make angles $\theta_1 = 53°$ and $\theta_2 = 47°$ to the horizontal. Find the coefficient of friction μ for which the masses move at constant velocity.

P85. A mass m is held at rest on an inclined plane, whose slope is α, by means of a horizontal force F (see Figure). The coefficient of static friction is μ_s. Find the

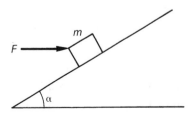

maximum F allowed before the body starts to move up the plane. Express your answer in terms of m, α, μ_s and g.

P86. A flatbed truck carries a box. The coefficient of static friction between the box and the truck is $\mu_s = 0.3$. What is the maximum acceleration the truck driver can allow so that the box does not slide? In the case where this maximum acceleration is just exceeded, find the distance the box travels *with respect to the truck* in the first 1s of the motion. Take the coefficient of sliding friction as $\mu = 0.2$.

P87. A computer monitor stands on a personal computer resting on a horizontal table. The monitor and computer have masses m, $M = 2m$ respectively. A student pulls the monitor horizontally with force F. The coefficients of friction between the computer and the table, and between the computer and the monitor are both μ. What is the maximum allowed force F_{max} such that the monitor does not move with respect to the computer? Will the computer move with respect to the table in this case? What happens if $F = 2F_{max}$? Justify your answer quantitatively.

P88. A book of mass M rests on a long table, with a piece of paper of mass $m = 0.1M$ in between. The coefficient of friction between all surfaces is $\mu = 0.1$. The paper is pulled with horizontal force P (see Figure). What is the minimum value of P required to cause any motion? With what force must the page be pulled in order to extract it from between the book and the table? Express your answers in units of Mg.

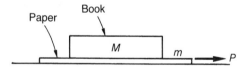

Paper   Book

M m P

■ WORK, ENERGY, AND POWER

P89. A child pushes a toy cart from rest on a smooth horizontal surface with a force $F = 5$ N, directed at an angle $\theta = 10°$ below the horizontal (see Figure). Calculate the work done by the child in 5 s if the cart's mass is $m = 5$ kg.

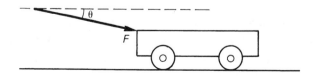

θ

F

P90. A train of mass $m = 1000$ metric tons accelerates from rest to a speed $v = 72$ km/h on a horizontal track. Calculate the work W done by the locomotive engine, neglecting friction.

P91. A bucket of water of mass $m = 10$ kg is raised from rest through a height of $h = 10$ m and placed on a platform. How much does its potential energy increase? What was the work done against gravity?

P92. A rollercoaster climbs to its maximum height $h_1 = 50$ m above ground, which it passes with speed $v_1 = 0.5$ m s^{-1}. It then rolls down to a minimum height $h_2 = 5$ m before climbing again to a height of $h_3 = 20$ m (see Figure). Neglecting friction, find the speed of the rollercoaster at these two points.

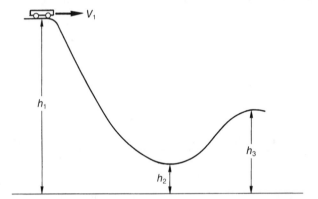

P93. A tennis player's serve gives the ball a kinetic energy $T_1 = 10$ J. Assuming that she serves from a height $h = 2$ m above the level of the court, find the speed with which the ball reaches the ground. Assume that the work done by the ball against air resistance is $W = 5$ J. (Mass m of a tennis ball $= 60$ g.)

P94. Show that the kinematic formula $v^2 = v_0^2 + 2ax$ for uniformly accelerated straight-line motion can also be derived from energy conservation.

P95. A high-jumper clears the bar at a height of $h = 2$ m with horizontal velocity $v_1 = 3$ m s^{-1}. Using conservation of energy, calculate the velocity with which he hits the landing platform (1 m above ground) and the direction of this impact velocity.

P96. An ambitious pole-vaulter wishes to clear a height $h = 6.10$ m. What is the minimum velocity he must reach on the runway? Explain why this is a minimum value.

P97. In order not to gain weight, a student of mass $m = 75$ kg wishes to get rid of the energy 1500 kJ gained in eating a hamburger by doing pushups. In each pushup he raises his center of mass through $h = 20$ cm. How many pushups must he perform? Is this result realistic?

P98. A water-skier is towed by a boat with horizontal force $F = 100$ N. She maintains a constant velocity $v = 36$ km/h. Find the work done against frictional forces, such as water and air resistance, in 10 s.

P99. Police drivers are taught that doubling the speed quadruples the braking distance. Why?

P100. A suitcase of mass $m = 20$ kg is dragged with a constant force $F = 150$ N along an airport ramp of slope $\alpha = 30°$ up to a height $h = 5$ m (see Figure). Find the coefficient μ of sliding friction if the suitcase's velocity is increased from zero at the bottom of the ramp to $v_2 = 1$ m s^{-1} at the height h.

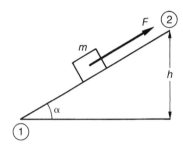

P101. Consider the pulley lifting arrangements of P27 and P28. Show that in each case the *total* work done by the woman in raising the mass M through a height h is the same, neglecting friction. Prove a similar result for the gear wheel arrangement of P29. Is the neglect of friction realistic in practice?

P102. A crane lifts a load of mass $m = 500$ kg vertically at constant speed $v = 2$ m s^{-1}. Find the power expended by the crane motor. What is the work done by the crane if the load is lifted through $h = 20$ m? A second crane is able to lift the same load at twice the vertical speed. Find the power expended and the work done in lifting the load through the same height.

P103. An electric pump draws water from a well of depth $d = 50$ m at a rate of 2 m^3 per second. The water is ejected from the pump with velocity $v_2 = 10$ m s^{-1}. What is the power consumption of the pump if its efficiency is $\eta = 0.8$ (80% efficiency)?

P104. A car of mass $M = 1000$ kg decelerates from a velocity $v = 100$ km/h to a stop in $t = 10$ s. At what average rate must the braking surfaces lose heat if their temperature is not to rise significantly?

P105. Animals of similar types but very different sizes tend all to be able to jump to roughly similar maximum heights (e.g. various types of dogs, or fleas and grasshoppers), although larger animals need more room to take off, roughly in proportion to their size. What does this suggest about the *rate* of energy release by muscles in larger animals compared with smaller ones?

P106. A mass m slides from rest at height h down a smooth curved surface which becomes horizontal at zero height (see Figure). A spring is fixed horizontally on the level part of the surface. Find the velocity of the mass immediately before encountering the spring, in terms of g, h. The spring constant is k. When the mass encounters the spring it compresses it by an amount $x = h/10$. Find k in terms of m, g, h. What height does the mass reach on returning to the curved part of the surface, if there are no energy losses in the spring?

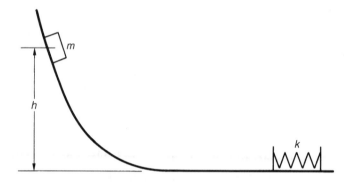

P107. A mass m is projected upwards with initial velocity v along an inclined plane of slope α, with $\sin \alpha = 1/\sqrt{2}$ (see Figure). The coefficient of sliding friction is $\mu = 0.1$. Using energy conservation, calculate the distance d the mass travels up the slope. Express your answer in terms of v, g. What must the minimum value of the coefficient of static friction μ_s be in order that the mass does not slide back? If μ_s is smaller than this value, with what velocity does the mass return to its starting point? Express your answer in terms of v.

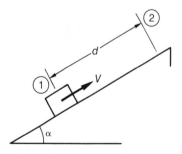

■ MOMENTUM AND IMPULSE

P108. A bird and an insect fly directly towards each other on a horizontal trajectory. The mass of the bird is M and that of the insect is m. The corresponding (constant) velocities are V, v. The bird swallows the insect and continues to glide in the same direction. Find its velocity U after swallowing the insect. Find U in terms of V in the case $m = 0.01M$ and $v = 10V$.

P109. A rifle has mass $M = 3$ kg and fires a bullet of mass $m = 10$ g with muzzle velocity $u = 700$ m s^{-1}. What is the recoil velocity v of the gun? From what height h would you have to drop the rifle on to your shoulder to feel the same kick?

P110. A rocket works by reacting against the momentum of its exhaust gases. Why are they often constructed with several stages?

P111. A cue ball has velocity u and collides head-on with a stationary pool ball of equal mass m on a smooth horizontal table. The collision is perfectly elastic (mechanical energy is conserved). What are the velocities v_1, v_2 of the two balls after the collision?

P112. In a one-dimensional collision, masses m_1, m_2 have velocities u_1, u_2 before the collision and v_1, v_2 afterwards. Show that if mechanical energy is conserved $v_2 - v_1 = -(u_2 - u_1)$, i.e. the bodies separate at the same speed they approached.

P113. An elementary particle of mass m_1 collides with a stationary proton of mass m_p. As a result of the collision the particle recoils along its direction of approach. A second elementary particle of mass m_2 continues to move forward after colliding with a proton. Give limits on the ratios $m_1/m_p, m_2/m_p$.

   If the velocity of the incoming particle is u in each case, find the final velocities of all the particles after the collisions in terms of u in the cases $m_1 = m_p/2, m_2 = 2m_p$.

P114. An elementary particle of mass m and velocity u collides with a stationary proton of mass m_p. Assuming that the total mechanical energy is conserved, calculate what fraction of the particle's energy is transferred to the proton.

P115. A mass m_1 moving with velocity u_1 collides with a stationary mass m_2. If the coefficient of restitution is $e (< 1)$, find the velocity v_2 of m_2 after the collision. Show that very little of the original kinetic energy is transferred to m_2 if m_1, m_2 are very different.

P116. If you want to knock a nail into the floor, why is it preferable to use a hammer than jump on the nail?

P117. A physicist observes the cue ball make a direct collision with a stationary pool ball and follow it with significant velocity. He concludes that the coefficient of restitution of pool balls is significantly smaller than 1. Is he correct?

P118. A baseball player swings the bat with velocity u_1 and hits a ball traveling with velocity u_2 (where $u_2 < 0$ of course) directly back towards the pitcher. If the bat and ball have masses m_1, m_2, with $m_1 \gg m_2$, and the collision is perfectly elastic, show that the ball leaves the bat with velocity at most $2u_1 - u_2$.

P119. A man sits at one end of a boxcar of internal length d, which is stationary on very smooth level rails. He tries to get the boxcar moving by throwing his boot, of mass m, at the opposite end with velocity u_1. Describe what happens, assuming the collision of the boot with the wall is completely inelastic (i.e. it does not rebound from the wall at all), and the total mass of the boxcar and man minus boot is M.

P120. In the previous question, what happens if instead of a boot the man throws a very bouncy ball, whose collision with the wall is completely elastic?

P121. A basketball player bounces the ball (coefficient of restitution e) so that it hits the floor vertically with velocity u_0. At that moment he falls over so that the ball bounces freely. If no other player intervenes, how high will the ball rise on the first bounce, and on the second bounce?

P122. In the previous question, how long does the player have to regain control of the ball before it stops bouncing?

P123. An artillery shell is fired at an angle $\theta = 45°$ to the horizontal with velocity $v_0 = 450 \, \text{m s}^{-1}$. At the maximum height of its trajectory the shell explodes, breaking into two parts of equal mass. One of these initially has zero velocity with respect to the ground. How far from the firing point does the other part fall back to the ground?

P124. A ball of mass $m = 0.1$ kg hits a rigid vertical wall at right angles with velocity $u = 20 \, \text{m s}^{-1}$. The impact is a height $h = 4.9$ m above the ground. It rebounds and falls to the ground a distance $x = 15$ m from the foot of the wall. What is the impulse exerted by the wall on the ball? Was the collision elastic?

P125. A bullet of mass $m = 10$ g is fired horizontally into a wooden block of mass $M = 7$ kg, which lies on a smooth horizontal table. The bullet is embedded in the block, and the block slides with velocity $V = 0.5 \, \text{m s}^{-1}$ after the impact. Find the muzzle velocity u of the gun firing the bullet, and the total mechanical energy lost in the impact.

P126. A wooden block of mass $M = 10$ kg hangs freely and at rest from vertical strings. A bullet of mass $m = 10$ g is fired into it and it rises by $h = 3$ cm.

What was the velocity u of the bullet? Where does most of its kinetic energy go?

P127. A dart of mass m is thrown horizontally with velocity u and sticks into a wooden block of mass $M = 8m$, which slides on a smooth horizontal table. The block's motion is resisted by an elastic spring with constant k (see Figure). Find the maximum distance through which the block compresses the spring. Express your answer in terms of m, u and k.

P128. A freight train moves steadily on a level track with velocity $v = 108$ km/h. Snow falls vertically on to it, and accumulates on it at a constant rate $r_m = 10 \, \text{kg s}^{-1}$. Calculate the additional power the locomotive must expend in order to maintain the train's speed despite the snow.

P129. A grain sack of mass $M = 10$ kg is dropped from a height of $h = 1$ m on to a platform. Calculate the impulse on the platform. Assume that the impact is short enough that gravity does not change the momentum during impact.

If the impact lasts $\Delta t = 0.1$ s, what is the *average* force on the platform during the impact?

P130. A steady stream of grain from a punctured sack falls vertically on a platform from a height $h = 1$ m. Each grain lands without bouncing, and 1000 grains land each second. Each grain has mass $m = 10$ g. What is the force on the platform, assuming again that gravity does not change the momentum during impact?

P131. A soccer goalkeeper of mass $m_g = 80$ kg punches a ball approaching him horizontally. The ball has mass $m_b = 0.5$ kg and velocity $u = 1 \, \text{m s}^{-1}$. Immediately after the punch the ball moves horizontally away along the direction of approach with velocity $v = 0.8u$. Assume that the impact lasts $\Delta t = 0.2$ s. What is the minimum value of the coefficient μ_s of static friction of the goalkeeper and the ground if he does not slide backwards?

P132. A boat and its occupant of total mass $M_b = 200$ kg contains 10 sacks of coal each of mass $m = 5$ kg. The boat is stationary because of engine failure. The occupant tries to reach land by throwing the sacks horizontally out of the boat. He throws each sack with a velocity v_r relative to the boat. Assuming no friction, what is the velocity after the first sack is thrown out? After the second sack is thrown out? Express your result in terms of v_r.

P133. Two cars of masses $m_1 = 1000 \, \text{kg}$ and $m_2 = 500 \, \text{kg}$, and velocities $u_1 = 18 \, \text{km/h}$ and $u_2 = 36 \, \text{km/h}$ collide at a right-angled intersection. After the collision they slide together as one. What direction (with respect to the first car's motion) do they move after the collision? With what velocity do they move? How much mechanical energy was lost on the collision?

P134. A cue ball hits a stationary pool ball of equal mass. After the collision the velocities of the balls make angles θ, ϕ to the original direction of motion of the cue ball. Find a relation between θ and ϕ, if the collision is regarded as elastic and the balls slide rather than rolling.

P135. A stationary spaceship of mass M is abandoned in space and must be destroyed by safety charges placed within it. The crew observe the explosion from a safe distance, and see that it breaks the ship into three pieces. All three pieces fly off in the same plane at angles $120°$ to each other. The velocities of the three fragments are measured to be $v, 2v$ and $3v$. What expressions will the crew find for the masses of the three fragments in terms of M? If all of the explosion energy E goes into the kinetic energy of the fragments, what was E in terms of M, v?

■ CIRCULAR AND HARMONIC MOTION

P136. A spaceship of mass $m = 10^4$ kg is in uniform circular motion $h = 200$ km above the surface of a planet of radius $R = 5000$ km. Each revolution takes $P = 2\,\text{h}$. Calculate the tangential velocity v of the spaceship, its angular velocity ω, and the centripetal force required to keep it in this orbit.

P137. A toy car of mass $m = 0.1$ kg is constrained to move in a circle of radius $r = 1$ m on a horizontal table by means of a string. Calculate the tension in the string if the car has constant angular velocity $\omega = 1$ rad s^{-1}.

P138. A plumbline hangs in equilibrium at latitude λ. Express the angle θ between the plumbline and the local vertical in terms of λ, and the Earth's radius, angular velocity and gravity R, ω, g. (Use the fact that $g \gg R\omega^2$ to simplify your answer.) Taking $R = 6400$ km, what is the maximum possible value of θ?

P139. A sports car attempts to take a bend which is an arc of a circle of radius $r = 100$ m. The road is horizontal and the car has constant speed $v = 80$ km/h. If the coefficient of static friction between the car tires and the road surface is $\mu_s = 0.4$, will the car stay on the road?

P140. A mass m is attached to a string and whirled in a vertical circle at constant speed. Calculate the difference between the tension at the lowest and highest points of the circle.

P141. A mass $m = 1$ kg is attached to a string and whirled in a vertical circle at constant speed. The radius of the circle is $r = 1$ m. What must the speed be to keep the string taut?

P142. In the arrangement described in P141 above, the string breaks when the mass is at its lowest point. In what direction and with what speed does the mass initially move?

P143. A mass M moves in a vertical circle at the end of a string of length L. Its velocity at the lowest point is v_0. Show that when the string makes an angle θ to the downward vertical its tension is

$$T = M\left(3g\cos\theta - 2g + \frac{v_0^2}{L}\right).$$

P144. A conical pendulum consists of a string of length $l = 2$ m and a bob of mass $m = 0.5$ kg. The pendulum rotates at a frequency $f = 2$ turns per second about the vertical. Calculate the tension T in the string and the angle α of the string to the vertical.

P145. An amusement park proprietor wishes to design a rollercoaster with a vertical circular loop in the track, of radius $R = 20$ m. Before the cars reach the loop, they descend from a maximum height h, at which they have zero velocity (see Figure). Assuming that the cars roll freely (no motor and no friction), how large must h be to keep the cars on the track?

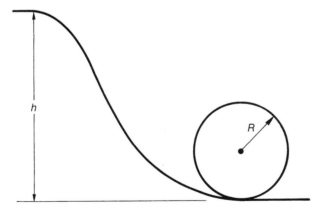

P146. A bobsleigh run consists of banked curves. One of the curves is circular and has radius $r = 10$ m, and is banked at an angle $\alpha = 60°$ to the horizontal. Neglecting friction, what is the maximum velocity at which a bobsleigh can take the curve?

P147. A fighter airplane has maximum level speed $v = Mc_s$, where M is the Mach number and $c_s \approx 340$ m s^{-1} is the speed of sound. The maximum acceleration the pilots can withstand without blacking out is $a = 6g$. How tight a turn can the fighter make at top speed if $M = 2$? What if $M = 3$?

P148. For the airplane of the previous question, what is the angle of banking to the horizontal in its tightest turns? If the pilot's mass is $m = 65$ kg, what is his apparent weight in the turns? (The lift on an airplane acts perpendicular to its wings.)

P149. A rail track has bends with radius of curvature as small as $r = 4$ km. If the passengers complain when accelerations exceed $a = 0.05g$, how fast can trains travel? Comment on the feasibility of trains running at $v = 400$ km/h.

P150. The dining car of a train uses water glasses of diameter $d = 8$ cm. If the maximum centripetal acceleration of the train is $a = 0.05g$, how close to the brim can these be filled without spilling?

(*Hint*: Remember that pressure = force per unit area, and consider the equilibria of the horizontal and vertical columns of water meeting at a point on the outer side of the glass.)

P151. Two equal masses m are attached by a string. One mass lies at radial distance r from the center of a horizontal turntable which rotates with constant angular velocity $\omega = 6$ rad s^{-1}, while the second hangs from the string inside the turntable's hollow spindle (see Figure). The coefficient of static friction between the turntable and the mass lying on it is $\mu_s = 0.5$. Find the maximum and minimum values r_{max}, r_{min} of r such that the mass lying on the turntable does not slide.

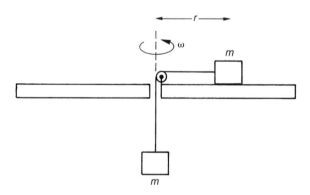

P152. The bends on a cycle track are semicircular, and the track is banked at an angle α to the horizontal. At what speed v_0 can a cycle and rider of mass M take these bends in horizontal circular motion of radius r even if there is no friction between the cycle tires and the track? Find the value of the frictional force f if the speed is $v_1 = 2v_0$, and also if it is $v_2 = v_0/2$. (Assume that the rider can always lean the cycle to avoid overturning.)

P153. A satellite is in a circular orbit whose height above the Earth is much less than the latter's radius $R_e = 6400$ km. What is its period?

P154. You are driving your car along a straight road at speed v_0 when you suddenly come to a T-intersection a distance r ahead with a river along the far side (see Figure). With maximum braking, the car would just stop without skidding with its nose overhanging the river bank. Should you attempt to take the turn?

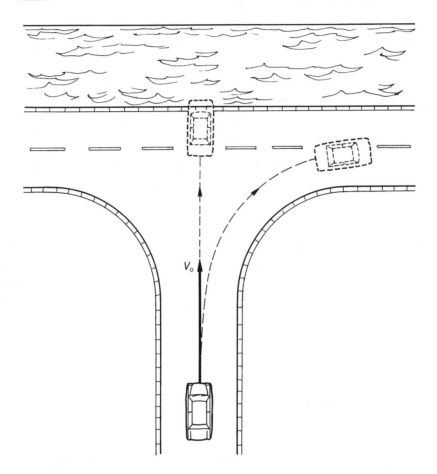

P155. A pendulum has length $l = 1$ m. How many swings (to the nearest whole number) does it perform in one hour?

P156. A pendulum is suspended from the ceiling of an elevator and set swinging while the elevator is at rest. A remote camera monitors the swing rate. How could you tell if the elevator moves up or down?

P157. When a mass $m = 1$ kg is hung vertically from a certain spring, it extends the spring by $\Delta x = 0.1$ m. Find the period of oscillation of the mass–spring system, if it lies on a smooth horizontal table.

P158. Two students have a spring (of unknown constant), two equal masses m and a string whose length can be adjusted. They wish to construct two oscillating devices (a mass–spring system and a pendulum) with exactly equal periods. What should they do?

P159. A mass $m = 0.2$ kg and a spring with constant $k = 0.5$ N m^{-1} lie on a smooth horizontal table. The mass is released a distance $x_0 = 0.1$ m from the equilibrium point. At what later time does the mass first pass through the point $x_1 = 0.02$ m from equilibrium? What is its velocity then?

P160. A pendulum of length $l = 9.8$ m hangs in equilibrium and is then given velocity $v_0 = 0.2$ m s^{-1} at its lowest point. What is the amplitude of the subsequent oscillation?

P161. A spring of constant $k = 0.5$ N m^{-1} and an attached mass m oscillate on a smooth horizontal table. When the mass is at position $x_1 = 0.1$ m its velocity is $v_1 = -1$ m s^{-1}, and at $x_2 = -0.2$ m it has velocity $v_2 = 0.5$ m s^{-1}. Find m and the amplitude A of the motion.

P162. A delicate piece of electronic equipment would be destroyed by vibration at frequencies greater than $\nu_m = 10$ s^{-1}. It is transported in a box supported by four springs. The total mass of the equipment and the box is $M = 5$ kg. What constant k would you recommend for the springs?

P163. A mass $M = 1$ kg is connected to two springs 1, 2 of constants $k_1 = 1$ N m^{-1}, $k_2 = 2$ N m^{-1} and slides on a smooth horizontal table (see Figure). In the equilibrium position it is given a velocity $v_1 = 0.5$ m s^{-1} towards spring 2. How long will it take to reach its maximum compression of spring 1? What will this be?

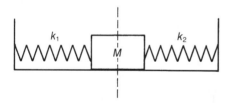

P164. In the previous question, how long does it take for the mass to reach the point where it compresses spring 1 by $x = -0.1$ m for the first time?

P165. When connected to a spring, a mass oscillates on a smooth horizontal table with period P. A second spring with the same constant is now connected between the first spring and the mass. What is the new oscillation period?

P166. A small platform of mass $m = 1$ kg lies on a smooth table and is attached to a wall by a spring. A block of mass $M = 4m$ lies on the platform. The platform–block system oscillates bodily with frequency $\nu = 1$ s^{-1} and amplitude $A = 0.1$ m. Find the spring constant k and the maximum horizontal force exerted on the block during the motion. If the coefficient of friction between the block and the platform is $\mu_s = 0.7$, how large an amplitude can the oscillation have without the block sliding from the platform?

■ GRAVITATION

P167. Compute the gravitational attraction force between the Sun and the Earth. (The mass of the Sun is 2×10^{30} kg, that of the Earth is 6×10^{24} kg, and their separation is $d = 1.5 \times 10^{11}$ m.)

P168. A planet has a circular orbit of radius a about the Sun, of mass $M_\odot$. What is the length P of the planet's year in terms of these quantities? (The planet's mass is much smaller than the Sun's.)

P169. The *effective gravity* g_{eff} at a point of the Earth's surface is defined by weighing an object and dividing the result by its known mass. What is the ratio of the effective gravity between the Earth's equator and the poles? (Assume the Earth is a sphere of mass $M_e = 6 \times 10^{24}$ kg and radius $R_e = 6.4 \times 10^6$ m.)

P170. What revolution period P_b must a spherical celestial body of mass M and radius R have if the effective gravity is zero at its equator? Find this value for the Earth (mass $M_e = 6 \times 10^{24}$ kg, radius $R_e = 6400$ km).

P171. Is it likely that a star can have a rotation period shorter than the value P_b defined in the previous question? The rotation periods of *pulsars* are detectable by radio astronomy and are found to be as short as $P_p = 5 \times 10^{-3}$ s. Are they more likely to be white dwarf stars (mass $M_w = 2 \times 10^{30}$ kg, radius $R_w = 5000$ km) or neutron stars (mass $M_n = 2 \times 10^{30}$ kg, radius $R_n = 10$ km)?

P172. A certain planet has mass M_p, which is twice the mass M_e of the Earth. On the planet the weight of any body is half the value it has on Earth. What is the planet's radius in terms of the Earth's radius R_e?

P173. The Earth's distance from the Sun is known to be $a = 1.5 \times 10^{11}$ m (the *astronomical unit*). Estimate the Sun's mass $M_\odot$.

P174. Estimate the mass M_e of the Earth from the facts that $g = 9.8$ m s^{-2} and $R_e = 6400$ km.

P175. A toy pistol uses a spring to fire a plastic bullet. On Earth the gun can propel the bullet to a maximum height h_e above the firing point. The gun is taken to the Moon and fired by an astronaut, who observes that the bullet can reach a height $h_m = 6h_e$. Find the acceleration g_m due to gravity on the Moon. (The heights h_e, h_m can be assumed much smaller than the radius of the Earth and Moon respectively, and air resistance is to be neglected.)

P176. An artificial satellite is called *geostationary* if it orbits directly over the equator at exactly the same angular velocity as the Earth. Find the height of such a satellite above the Earth. (Earth's mass $M_e = 6 \times 10^{24}$ kg, radius $R_e = 6.4 \times 10^6$ m.)

P177. Clearly, it would be useful to have a geostationary communications satellite placed directly over every large city. Yet there are none. Why not?

P178. A space shuttle is in a circular orbit at a height H above the Earth. A small satellite is held above the shuttle (i.e. directly away from the Earth) by means of a rod of length h and then released. What is its initial motion relative to the shuttle?

P179. The space shuttle of the previous question fires a retro rocket, i.e. one directed with its exhaust pointing forward. What will happen to the shuttle?

P180. An artificial satellite is in a circular orbit of radius r about a planet of mass M. Find its speed and angular momentum per unit mass. The planet's atmosphere exerts a drag on the satellite in such a way that its orbit remains circular. Does it slow down or speed up?

P181. Show that the Sun's gravitational pull on the Moon is more than twice as large as the Earth's. Why does the Moon not fly off? (Mass of Sun $M_\odot = 2 \times 10^{30}$ kg, mass of Earth $M_e = 6 \times 10^{24}$ kg; Sun–Earth distance $a = 1.5 \times 10^{11}$ m, Earth–Moon distance $r = 3.9 \times 10^8$ m.)

P182. A non-rotating planet of radius R has a circular orbit of radius a about the Sun (mass M). Show that on the planet's surface, the *effective inward* gravitational acceleration g_{eff} is lowest at the points nearest to and furthest from the Sun, and highest on the circle equidistant from these two points (see Figure). Assuming $a \gg R$, show that the difference in accelerations is approximately $3GMR/a^3$.

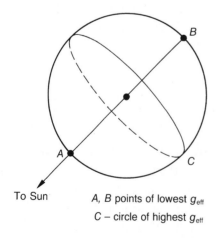

To Sun   A, B points of lowest g_{eff}

C – circle of highest g_{eff}

P183. If in the previous problem the planet is completely covered by an ocean, what is the ratio of its maximum to minimum depths? If the planet rotates, what would the inhabitant of a small island observe?

P184. Show that the Moon raises about twice the tide that the Sun does. When would you expect the maximum and minimum tides to occur? (Masses $M_\odot, M_m$ of Sun and Moon are $2 \times 10^{30}, 7 \times 10^{22}$ kg; Earth–Sun distance $a = 1.5 \times 10^{11}$ m, Earth–Moon distance $b = 3.8 \times 10^8$ m.)

P185. Why are no tides observed on the Great Lakes or the Mediterranean?

P186. Because the Earth does not rotate synchronously with the Moon, dissipation in tides cause angular momentum to be transferred from the Earth's spin to the Moon's orbit. Show that the Earth–Moon distance and the length of the Earth day must be (slowly) increasing. If the process will stop when the Earth–Moon distance is about 1.5 times its current value, what will the length of the day be? (Earth's mass $M_e = 6 \times 10^{24}$ kg, current Earth–Moon distance $b = 3.8 \times 10^8$ m.)

P187. What is the escape velocity from Earth? (i.e. the velocity with which an object must be launched in order to escape to infinity). (Earth mass $M_e = 6 \times 10^{24}$ kg, Earth radius $R_e = 6400$ km.)

P188. How does the escape velocity from Saturn compare with that from Earth (compare P187)? (Saturn mass $M_s = 95 M_e$, Saturn radius $R_s = 9.4 R_e$.)

P189. A space probe is launched, but by mishap achieves a vertical speed v_0 only three-quarters of the escape velocity. It then goes into a circular orbit: find its radius in terms of the Earth's radius R_e.

P190. A rocket is launched from Earth (mass M_e, radius R_e) with velocity v_0, and reaches radial distance $r = 6R_e$ with velocity $v = v_0/10$. Express v_0 in terms of M_e, R_e.

P191. What is the maximum height that the rocket of the previous problem could reach if launched vertically?

P192. A space station orbits the Earth (radius R_e) at height $R_e/2$ above its surface. What is its speed? The astronauts on board launch a rocket. What minimum speed with respect to the station does it need in order to leave the Earth's gravitational field?

P193. The escape velocity from a black hole of mass M equals the speed of light c. What is its radius? Evaluate this if (a) $M = $ Sun's mass $M_\odot$, (b) $M = 3M_\odot$. ($M_\odot = 2 \times 10^{30}$ kg.)

P194. Consider the $3M_\odot$ black hole of the previous question. How does its average density compare with that of the atomic nucleus? ($\rho_{nuc} \approx 10^{18}$ kg m^{-3}.)

P195. The nuclei of some galaxies are thought to contain supermassive black holes with $M = 3 \times 10^9 M_\odot$. How do their average densities compare with that of air? ($\rho_{air} = 1.3$ kg m^{-3}.)

■ RIGID BODY MOTION

P196. A car accelerates uniformly from rest for 10 s, when its velocity is $v = 10$ m s^{-1}. Assuming that the wheels do not slip, find the final angular velocity ω of the wheels and the angular acceleration α. The radius of the wheels is $R = 0.5$ m.

P197. Four masses are attached to a massless circular hoop of radius $R = 1$ m as shown in the Figure. Find the moment of inertia of the resulting configuration about a perpendicular (z) axis through the hoop's center ($m_1 = 1$ kg, $m_2 = 2$ kg, $m_3 = 3$ kg). A force $F = 5$ N is applied tangentially to the rim of the hoop. What is its angular acceleration α?

P198. In the previous problem, what are the moments of inertia I_x, I_y about the x and y axes respectively?

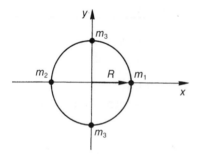

P199. A uniform circular cylinder of mass m, radius r and length $l = r$ is allowed to roll horizontally down an inclined plane of angle $\alpha = 60°$ to the horizontal (see Figure). It starts from rest with its center of mass at a height $h + r$ above

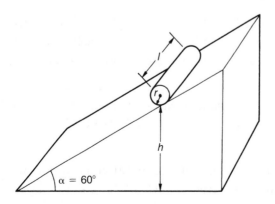

the base of the plane. Calculate the time t_c for it to reach the bottom (i.e. to roll through a height h). Compare your result with the corresponding time t_s for a uniform sphere of mass m and radius r. Assume that there is no slipping in either case. Compare t_c, t_s with the time t_0 for a mass to slide through the same height without friction.

P200. A solid uniform cylinder of mass m and radius r rolls without slipping down an inclined plane with a vertical circular loop of radius R fixed at the bottom (see Figure). The cylinder starts to roll from rest at height h. You may assume that $r \ll h, r \ll R$. What is the minimum value h_m of h such that the cylinder does not fall from the circular loop? A cylinder with the same mass m all concentrated in a thin shell at radius r is released from rest at $h = h_m$. Does this cylinder complete the loop or not?

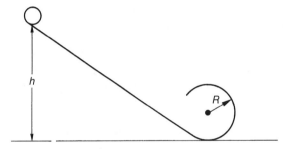

P201. A body of mass M has moment of inertia I about an axis through its center of mass. Show that its moment of inertia about a parallel axis a distance d from the first is $I + Md^2$ (*parallel axes theorem*).

P202. A mass m hangs from a string whose other end is wound on a circular pulley of mass $M = 2m$ and radius R. The string does not stretch or slip. Find the linear acceleration a and the string tension T in terms of m, g, and R. If the mass starts from rest, calculate the total angular momentum L about the pulley's center after the mass has descended a height $h = R$.

P203. A child's top is given angular momentum L about a vertical axis. Why does it not fall over until this has been lost? Explain qualitatively what happens if one tries to push over a spinning top.

P204. A rifle barrel has a spiral groove which imparts spin to the bullet. Why?

P205. A turntable consists of a thin horizontal disc of mass M and radius R, and rotates without friction at constant angular speed ω. At a certain instant a drop of glue of mass $m = M/10$ falls vertically on to the turntable and adheres to a point at a distance $r = 3R/4$ from the axis. Find the new angular velocity of the turntable.

P206. A pendulum consists of a uniform rod AB of length $l = 0.5$ m and mass $M = 1$ kg. Calculate the period P of the pendulum in the cases
   (a) – the pendulum is suspended from point A,
   (b) – it is suspended from a point C such that $AC = l_C = l/4$.

P207. A skater spins with angular velocity $\omega_b = 6$ rad s^{-1} with his arms extended. How fast will he spin with his arms by his sides?
   Treat the skater's body as a uniform cylinder of radius $R = 20$ cm; approximate his arms as uniform rods of length $L = 70$ cm and mass $m = 4.5$ kg. His total mass excluding arms is $M = 70$ kg.

P208. A man of mass $m = 80$ kg stands on a flat horizontal disk of mass $M = 160$ kg near its edge, at radius $r = 2.5$ m. The disk is free to rotate about its axis. At a certain instant the man begins to walk around the disk edge with constant velocity $v = 2$ m s^{-1} with respect to the Earth. If his feet do not slip on the disk, how long will it take the man to return to the same point on the disk? What will happen if the man stops walking?

P209. A pool ball of mass m and radius R is given an initial sliding velocity v_0 (no rotation) on a horizontal pool table. The coefficient of friction between the ball and the table is μ. How long will it take for the ball to start a pure rolling motion (no sliding)? What will be its velocity v at that point?

P210. A baseball player strikes the ball a distance x from the handle of the bat, which has mass M and moment of inertia I about the center of mass. If the latter lies a distance l from the handle, how should the player choose x so that his hands experience no reaction force?

P211. A pool ball has radius l and mass M. A player hits it a horizontal blow with her cue at height h above the table. How should she choose h so that the ball rolls without sliding?

P212. In P208, if there is friction about the disk axis, what happens when the man stops walking?

CHAPTER TWO

ELECTRICITY AND MAGNETISM

■ SUMMARY OF THEORY

1. Coulomb's Law

● The force between two charges q_1, q_2 with separation r is

$$F = \frac{q_1 q_2}{4\pi\epsilon_0 r^2}.$$ (1)

in vacuo (or air), where the charges are in coulombs (C). The force acts along the line joining the charges, and is repulsive for charges of the same sign and attractive for charges of opposite sign. ϵ_0 is a constant, the permeability of vacuum.

2. Electric Field

● We define the *electric field E* as the force on per unit static positive charge. The units are N C^{-1}. A general charge q experiences force qE in the same direction as E if $q > 0$, and the opposite direction otherwise. The electric field due to a point charge q is

$$E = \frac{q}{4\pi\epsilon_0 r^2},$$ (2)

and is radial. If certain charge distributions produce electric fields $E_1, E_2, \ldots$ at a point, the resultant electric field has components

$$E_x = E_{1x} + E_{2x}\ldots$$ (3)

and similarly for the other components E_y, E_z.

47

The electric charge and electric field vanish everywhere inside a perfect conductor: all charge must be confined to a thin layer at the surface.

● *Gauss's law* states that the flux of electric field over a closed surface is $1/\epsilon_0$ times the total charge enclosed. This agrees with (1) for a point charge, and shows that for example

$$E = \frac{\lambda}{2\pi\epsilon_0 r} \tag{4}$$

at distance r from a very long line of charge, distributed at λ C m^{-1}.

3. Potential

● The *potential* at a point is the work done against electric forces in moving unit positive charge from infinity to the point. The units are volts = J C^{-1}. The work done in moving a charge from one point to another depends only on the potential difference between the points, and not on the path between them. The potential difference in a uniform field E between two points is

$$V = Ez, \tag{5}$$

where z is the distance measured in the direction of the field. The potential at distance r from a point charge q is

$$V = \frac{q}{4\pi\epsilon_0 r}. \tag{6}$$

Inside a perfect conductor the potential is constant, since the field vanishes.

4. Capacitance

● A *capacitor* is a device for storing charge, consisting of conductors surrounded by an insulator or dielectric. The *capacitance* C of a capacitor is a measure of its ability to store charge and is defined as

$$C = \frac{|q|}{|\Delta V|}, \tag{7}$$

where q is the charge on either conductor and ΔV is the potential difference causing the accumulation of this charge.

● The capacitance of a *parallel plate capacitor* is

$$C = K_d \epsilon_0 \frac{A}{d}, \tag{8}$$

where K_d is a dimensionless constant characteristic of the insulator between the plates (the dielectric constant), A is the area of one plate, and d the plate separation. It is assumed that $A \gg d^2$.

● If capacitances $C_1, C_2, \ldots$ are connected in series, the total capacitance is C, where

$$\frac{1}{C} = \frac{1}{C_1} + \frac{1}{C_2} + \ldots \tag{9}$$

If they are connected in parallel the total capacitance is

$$C = C_1 + C_2 + \ldots \tag{10}$$

● The electrostatic energy stored in a capacitor is

$$U = \frac{CV^2}{2} = \frac{Vq}{2} = \frac{q^2}{2C}. \tag{11}$$

5. Current and Resistance

● Electric *current* is defined as (charge transported)/(time). The *electromotive force*, usually abbreviated to emf, of a battery is equal to the potential difference (or voltage drop) between its terminals when no current flows.

● The *resistance R* of part of an electric circuit is defined as the potential difference required to make unit current flow. It is measured in ohms (Ω). The voltage required to make current I flow is thus

$$V = IR, \tag{12}$$

which is known as *Ohm's law*.

● The *resistivity ρ* of a medium is defined as

$$\rho = \frac{RA}{l}, \tag{13}$$

where R is the resistance of a length l of a cylinder of cross-sectional area A made of the medium. ρ is measured in Ω m.

● The *power dissipated in a resistor* is

$$P = VI = I^2 R = \frac{V^2}{R}, \tag{14}$$

which is lost as heat.

● If resistors $R_1, R_2, \ldots$ are connected in series the total resistance is

$$R = R_1 + R_2 + \ldots, \tag{15}$$

while if they are connected in parallel the total resistance is R, where

$$\frac{1}{R} = \frac{1}{R_1} + \frac{1}{R_2} + \ldots \tag{16}$$

● The flow of current in a *direct current (DC) circuit* is determined by *Kirchhoff's laws*. These state that:

(a) – The total net current at each junction of a circuit is zero.

(b) – The total potential drop around any closed circuit is zero.

Note that in (a), currents are counted as having opposite signs when flowing into and away from the junction. In (b) we must be careful to include all the potential drops $V = IR$ caused by resistors, as well as any emf sources.

6. Magnetic Forces and Fields

● A magnetic field is present if a charge experiences a force resulting from its motion. The *magnetic force F* on a charge q moving with velocity v at angle θ to the field direction is

$$F = qvB \sin \theta, \tag{17}$$

where the direction of F is given by the *right-hand rule*: point the extended fingers of the right hand in the direction of the field and the thumb in the direction of motion of the charge. The palm then pushes in the direction of the magnetic force on a positive charge. The force is reversed if the charge is negative. The unit of magnetic field is the *tesla* (T), sometimes called the *weber per square meter*. The Earth's magnetic field is of the order 10^{-4} T. The total force on a charge due to both electric and magnetic fields is usually called the *Lorentz force*.

The force on a short length Δl of wire carrying current I is

$$\Delta F = IB\Delta l \sin \theta, \tag{18}$$

with the direction given as before. The force exerted by uniform field B on any length l of a straight wire is

$$F = IBl. \tag{19}$$

● A magnet of dipole moment μ placed at angle θ to the direction of a magnetic field B will experience a torque

$$\Gamma = -\mu B \sin \theta \tag{20}$$

trying to align it to the field direction.

● All magnetic fields result from electric currents. The fields of permanent magnets are caused by charge motions at a microscopic level.

Ampère's law states that the sum of the products of the tangential magnetic field with the length of each element of a closed curve is μ_0 times the total current enclosed by the curve. μ_0 is a constant, the permittivity of vacuum.

● The field of a long straight wire carrying current I is

$$B = \frac{\mu_0 I}{2\pi r}, \tag{21}$$

at distance r from the wire. The fieldlines are circles centered on the wire with planes perpendicular to it.

The field inside a long *solenoid* with n loops per unit length carrying current I has the constant value

$$B = \mu_0 n I \tag{22}$$

in the interior.

The field inside a toroidal coil with N loops carrying current I is

$$B = \frac{\mu_0 N I}{2\pi r}, \tag{23}$$

where r is the radial distance of the point from the center of the torus.

● The magnetic force per unit length between two long parallel wires with separation d carrying currents I_1, I_2 is

$$F_m = \mu_0 \frac{I_1 I_2}{2\pi d}. \tag{24}$$

The force is attractive if I_1 and I_2 are in the same direction and repulsive otherwise.

7. Electromagnetic Induction

● The magnetic flux Φ through a surface of area A is defined as

$$\Phi = BA \cos \theta, \tag{25}$$

where θ is the angle between the normal to the surface and the field direction, and it is assumed that B and θ do not vary appreciably over the surface.

● *Faraday's law of magnetic induction* states that the rate of change of magnetic flux through a circuit is minus the induced emf in the circuit, i.e.

$$\mathcal{E} = -\frac{\Delta\Phi}{\Delta t},\tag{26}$$

where $\Delta\Phi, \Delta t$ are the changes in flux and time.

The minus sign in this equation expresses what is sometimes called *Lenz's law*: the induced emf is always in the direction opposing the change in magnetic flux that produced it.

As a corollary, one can show that the emf induced between the ends of a rod of length l moving with uniform velocity v perpendicular to itself and at angle θ to the field is

$$\mathcal{E} = Blv\sin\theta\tag{27}$$

The direction of the emf is given by the right-hand rule.

● A time-varying current in a circuit induces an emf. This effect is called *self-inductance*. If a change ΔI in time Δt induces emf V, we may write

$$V = -L\frac{\Delta I}{\Delta t}.\tag{28}$$

The minus sign here again reflects Lenz's law. The coefficient L is determined by the geometry of the circuit and is called its *self-inductance*. The units of L are henries (H).

The self-inductance of a coil of N turns, cross-sectional area A generating magnetic field B from current I is

$$L = \frac{NBA}{I}.\tag{29}$$

■ ELECTRIC FORCES AND FIELDS

P213. Two charges $q_1 = 2 \times 10^{-5}$ C and $q_2 = 4 \times 10^{-5}$ C are held a distance $d = 1$ m apart. Calculate the force exerted by these two charges on a charge $Q = 10^{-5}$ C, if it is placed halfway between them. Is there a point between the two charges where the force vanishes?

P214. Charges $q_1 = 0.09$ C, $q_2 = 0.01$ C are a distance $l = 1$ m apart. A charge Q is held fixed on the line between them, a distance x from q_1. What value must Q, x have for q_1, q_2 to feel no net force?

P215. A charge $Q = 1\,\mathrm{C}$ is at the origin of coordinates (see Figure). Calculate the magnitude and direction of the force exerted on it by the charges $q_1 = -0.5 \times 10^{-6}\,\mathrm{C}$ at position $(0, 3)$, and $q_2 = 10^{-6}\,\mathrm{C}$ at position $(4, 0)$, where all distances are in meters.

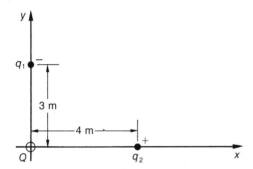

P216. Charges $q_1 = -2 \times 10^{-6}\,\mathrm{C}$ and $q_2 = 3 \times 10^{-6}\,\mathrm{C}$ are fixed at the points $A_1(8,0)$ and $A_2(0,10)$ respectively in a Cartesian coordinate system, with the length units being centimeters. Calculate the force on a charge $q_3 = -10^{-6}\,\mathrm{C}$ placed at the origin.

P217. A small sphere carries charge Q and can slide freely on a horizontal insulating rod of length l. Two further small spheres have charges $q, 4q$ and are fixed to the ends of the rod. Where does the sliding sphere come to rest?

P218. Charges q_1, q_2, q_3, q_4 are placed at the corners of a square of side $a = 2$ m. If $q_1 = q_2 = q_3 = Q = 1$ C and $q_4 = -Q$, find the electric field at the center of the square.

P219. In a hydrogen atom the electron is at a distance $a = 5.28 \times 10^{-11}$ m from the nucleus, which consists of a single proton. What is the electric field of the nucleus at the position of the electron? What is the force on the electron? If the electron is in a uniform circular orbit around the nucleus what are its speed and orbital period? (Treat the electron's motion using classical mechanics.)

P220. The electric field just above the Earth's surface is known to be $E_e = 130\,\mathrm{N}\,\mathrm{C}^{-1}$. Assuming that this field results from a spherically symmetrical charge distribution over the Earth, find the total charge Q_e on the Earth. (Earth's radius $R_e = 6400$ km.)

P221. Assuming that the Earth's field mentioned in the last problem acts vertically, what charge q would a ball of mass $m = 10$ g have to have to hover in mid-air?

P222. Point charges q and $9q$ are a distance l apart. Where should a third charge Q be placed so that the net force on all three charges vanishes? What is the required value of Q?

P223. Two horizontal plates of opposite charge create a constant electric field $E_0 = 1000$ N C^{-1} directed vertically downwards (see Figure). An electron of mass m_e and charge $-e$ is fired horizontally with velocity $v_0 = 0.1c$ between the plates. Calculate the electron's acceleration; if the plates have length $l_0 = 1$ m, find the electron's deflection from the horizontal when it emerges. Neglect gravity in this calculation: is this justified?

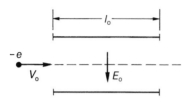

P224. A beam of electrons is injected horizontally with velocity $v_e = 10^6$ m s^{-1} into a vacuum tube in which there is a constant electric field $E_0 = 2000$ N C^{-1} directed vertically upwards. At the end of the tube the beam hits a fluorescent screen $h = 10$ cm lower than the injection point.
(a)   If the polarity of the field is reversed what happens to the impact point?
(b)   What is the horizontal distance l between the injection point and the screen?

P225. In the cathode ray tube of a television set electrons are accelerated by a high voltage V. They are then deflected by a pair of horizontal plates of separation d, length l and potential difference V_P (see Figure). The electrons then hit a fluorescent screen at distance L from the plates. How must V_P be chosen so that the electrons just clear the plates? (Neglect gravity.)

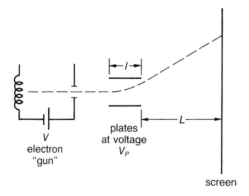

P226. In an experiment to measure the electron charge $-e$ (a modern version of Millikan's oil drop experiment) plastic balls of radius $r = 10^{-6}$ cm and density $\rho = 0.8$ g cm^{-3} are placed in vacuum between two horizontal charged plates, which create a uniform electric field E, directed vertically downwards.

The field is gradually adjusted until some balls remain stationary. In one experiment, balls were found to remain stationary for fields $E_1 = 3.13 \times 10^4 \, \text{N C}^{-1}$, $E_2 = 3.69 \times 10^4 \, \text{N C}^{-1}$. Assuming that the balls' charges differ by exactly one electron, estimate e.

P227. Two masses $m = 1$ kg with equal charges Q are suspended by light strings of length $l_0 = 1$ m from a point. The strings hang at $30°$ to the vertical; what is Q?

P228. Two small metal balls are tied together by a taut string of length $d = 1$ m. The balls are electrically neutral and the string can withstand a maximum tension $T_{\text{max}} = 1000$ N. Calculate how many electrons would have to be added in equal numbers to each ball before the string breaks. Is this a large number compared to the number in a metal ball of mass 10 g?

P229. Two alpha particles (helium nuclei, charge $q_\alpha = 2e = 3.2 \times 10^{-19}$ C, mass $m_\alpha = 6.68 \times 10^{-27}$ kg) are a distance $d = 2 \times 10^{-14}$ m apart. Calculate their electrostatic repulsion. How does this force compare with their gravitational attraction?

P230. What electric field E_0 is required to exert a force on an electron equal to its weight on Earth? Compare this field with that produced by a proton at a distance of $a_0 = 10^{-10}$ m ($a_0 \sim$ typical size of an atom).

P231. A very long solid cylinder has radius $R = 0.1$ m and uniform charge density $\rho_0 = 10^{-3}$ C m^{-3}. Find the electric field at distance r from the axis inside the cylinder in terms of r/R.

P232. A charge q of mass m is constrained to move along the y-axis. Charges $Q = -q/2$ are placed on the x-axis at positions $x = \pm a$. Calculate the force on the charge q at any position y. Show that the origin is an equilibrium point. Prove that for $y \ll a$ the charge will oscillate about the origin. Find the period of this oscillation if $q = 10^{-2}$ C, $m = 1$ kg and $a = 2$ m.

P233. Electric charge is distributed at a line density $\lambda = -2$ C m^{-1} along an infinite line. A point charge $q = 0.01$ C of mass $m = 1$ kg orbits in a circle whose plane is perpendicular to the line. What is its velocity?

P234. Point charges q and $-q$ are located at points $A(0, -a)$ and $B(0, a)$ in a cartesian coordinate system (this type of arrangement is known as an electric dipole). Find the electric field at any point on the x-axis. Show that for $x \gg a$ the field decays as x^{-3}.

P235. A large square insulating plate of side a and negligible thickness is uniformly charged with total charge $100Q$. The plate is placed in the y–z plane. A spherical shell of radius r is uniformly charged with total charge Q and has its center at the point $(d, 0, 0)$ (see Figure). If $a = 100d$ and $r = d/5$, calculate

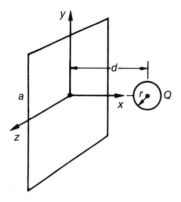

the electric field at any point P_1 inside the shell, and at the point $P_2 = (d/2, d/2, 0)$. Express your answer in terms of Q, ϵ_0 and d.

P236. A uniformly charged insulating sphere of radius a is surrounded by a concentric conducting shell of inner and outer radii $2a, 3a$. The total charge of the conducting shell is zero and that of the insulating sphere is Q. Find the electric field at all points. Plot your result.

P237. A point charge q is at the center of a thin spherical shell of radius R carrying uniformly distributed charge $-2q$. A second concentric shell of radius $2R$ has uniformly distributed charge $+q$. Find the electric field $E(r)$ for all values of the radial coordinate r, and plot your results schematically.

P238. A long coaxial cable consists of a uniform cylindrical core of radius R with uniform volume charge density ρ and a hollow cylindrical sheath of outer radius $2R$ with *surface* charge density σ (see Figure). What value must σ take (in terms of ρ, R) so that the external electric field vanishes?

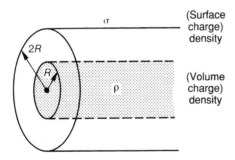

P239. A very long cylinder of radius R has uniform charge density ρ C m^{-3}. Find the magnitude and direction of the electric field E everywhere. Plot E as a function of r, the distance from the axis of the cylinder.

P240. A point charge q of mass m is released from rest at a distance d from an infinite plane layer of surface charge $\sigma = -q/d^2$. The point charge can pass through the layer without disturbing it. Find the acceleration and velocity of the charge as a function of position. Show that the motion is periodic and find the period P.

ELECTROSTATIC POTENTIAL AND CAPACITANCE

P241. Two charges $q_1 = 5 \times 10^{-8}$ C and $q_2 = -5 \times 10^{-8}$ C are held at a distance of $d = 12$ m. Calculate the electrostatic potential at the points A and B in the Figure.

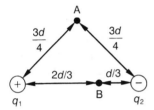

P242. In order to hold a small charged body in equilibrium against gravity an electric field $E = 2 \times 10^4 \, \mathrm{N\,C^{-1}}$ is needed. What potential difference would be required between two plates held $d = 2$ cm apart in order to achieve this field?

P243. An elementary particle of charge $q = +e$ and mass $m = 2m_p$ (m_p is the proton mass) falls from rest at infinity towards the Earth, assumed electrically neutral. Find its kinetic energy T when it reaches a height $h = 100$ km above the Earth's surface. (Mass M_e of earth $= 6 \times 10^{24}$ kg, radius $R_e = 6400$ km.)

The same particle is now projected from infinity towards the Earth with the kinetic energy T found above. What must the total charge Q_e on the Earth be if the particle never reaches its surface?

P244. An elementary particle of mass m and charge $+e$ is projected with velocity v at a much more massive particle of charge Ze, where $Z > 0$. What is the closest possible approach distance b of the incident particle?

P245. Two particles with electric charges $q_1 = +2e$ and $q_2 = -e$ have masses $m_1 = 4m_p$ and $m_2 = m_p$ respectively. ($-e$ is the electron charge and m_p the proton mass.) The particles are released from rest when very far apart, and approach each other under their mutual electrostatic attraction. Find their relative velocity when they are at a distance $L = 10^{-9}$ m apart.

P246. An electron volt (eV) is an energy unit equal to the kinetic energy acquired by an electron accelerated through a potential difference of 1 volt. This is a common energy unit in atomic and nuclear physics. Express the unit in joules, given that the electron charge is $e = -1.6 \times 10^{-19}$ C. What potential difference is required to accelerate an alpha particle (charge $+2e$) to an energy of 10^5 eV?

P247. Charges $q_1 = 10^{-6}$ C, $q_2 = 2 \times 10^{-6}$ C and $q_3 = -3 \times 10^{-6}$ C are held at the points $(x_1 = 0, y_1 = 0), (x_2 = 3, y_2 = 0), (x_3 = 1, y_3 = 4)$ of a Cartesian coordinate system, the units of length being meters. Calculate the potential at the point P with coordinates $(2, 2)$.

P248. A uniform electric field $E_0 = 100$ N C^{-1} in the positive y-direction (see Figure) is maintained between the planes $y = 0$ and $y = y_1 = 5$ cm. What is the potential difference ΔV between the two planes? A charge $Q_0 = 1$ C is moved quasistatically from the upper plane [position $(0, y_1)$] along the y-axis to the lower plane, i.e. to $(0, 0)$. What is the mechanical work done? Show explicitly that the same work is done if the charge is brought to the lower plane along a diagonal path to the point $(x_1, 0)$, where $x_1 = 5$ cm (see Figure).

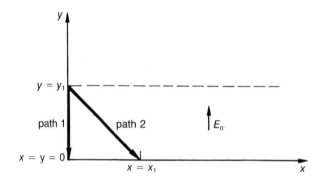

P249. The electric potential at a certain distance from a point charge is 500 volts. The electric field at that point is $100\,\mathrm{N\,C^{-1}}$. What is the value Q_0 of the charge, and what is the distance of the point from the charge?

P250. Two points A and B lie a distance $d = 10$ m apart in the direction of a uniform electric field $E = 200$ N C^{-1}. What is the potential difference between A and B? What work is done moving a charge $q = -0.01$ C from A to B
    (a) – directly along the straight line AB; and
    (b) – by moving 1 m from A to the left of the line, and then directly towards B in a straight line?

P251. A spherical conducting shell of radius $a = 10$ m is charged by attaching it to a DC source of voltage $\mathcal{E} = 1000$ V. What is its final charge? How much work is done in bringing a test charge $q = 1$ μC from infinity to the surface of the shell? If the test charge can penetrate the shell, is extra work required to bring it to the center?

P252. $N = 1000$ spherical drops of mercury (which can be regarded as a perfect conductor) each of radius r all have the same potential V when they are far apart. They merge and form one spherical drop. Find the original charge on each drop, the charge Q on the merged drop, and its potential V_1. (Express your results in terms of r, V and physical constants.) How and why does the total electrostatic energy change in the merging?

P253. In a Rutherford scattering experiment a beam of alpha particles, each with charge $q_\alpha = 4e$ and energy $E_\alpha = 1$ MeV $= 10^6$ eV is incident on a gold foil. See P246 for the definition of an electron volt (eV). What is the distance of closest possible approach d of an alpha particle to a gold nucleus (charge $q_{\mathrm{Au}} = 79e$)? What is the ratio of an alpha particle's kinetic energy T_α and its electric potential energy U when it is a distance $2d$ from a gold nucleus?

P254. An electron is accelerated through a potential difference of 1000 V, thus acquiring kinetic energy $E_e = 1000$ eV $= 1$ keV (see P246). What is its velocity? If $n = 10^{10}$, such electrons hit an electrode every second. What is the force on the electrode? What is the force if the electrons are replaced by protons of energy 1 keV?

P255. An accelerator creates an electron beam equivalent to a current of $I = 10^{-4}$ A and energy $E_e = 10^{10}$ eV per electron. How many electrons would hit a target in 1 s, and how much energy would be deposited?

P256. A parallel plate capacitor of capacitance $C = 10^{-8}$ F is connected through a resistor R to a power supply $\mathcal{E} = 1000$ volts. What charge Q accumulates on each plate? What is the energy thereby stored in the capacitor? When the capacitor is fully charged it is disconnected from the circuit and the distance between its plates is doubled. What is the stored energy now? Where did the extra energy come from?

P257. To measure the capacitance of an electrometer it is first charged to a potential $V_0 = 1350$ V. It is then connected by a conducting wire to a distant metal sphere of radius $r = 3$ cm. As a result the electrometer's potential drops to $V_1 = 900$ V. What is the capacitance C of the electrometer, and the charges Q, Q_1 on it before and after connecting it to the sphere?

P258. In the circuit shown in the Figure, the capacitance C_1 has the value 8 μF. The space between the plates of C_2 is filled with material of dielectric constant

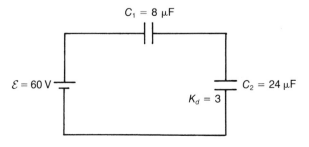

$K_d = 3$, and as a result $C_2 = 24~\mu F$. Calculate the potential differences V_1, V_2 across the capacitors, and the *total* electrostatic energy stored in them. Recalculate these quantities if the dielectric material is removed from C_2.

P259. Two capacitors C_1 and $C_2 = 2C_1$ are connected in a circuit with a switch between them (see Figure). Initially the switch is open and C_1 holds charge Q. The switch is closed and the system relaxes to a steady state. Find the potential V, electrostatic energy U and charge for each capacitor. Compare the total electrostatic energy before and after closing the switch, expressed in terms of C_1 and Q.

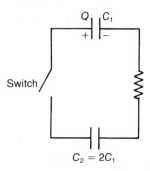

P260. A capacitor has parallel square conducting plates of side l a distance $d = l/100$ apart (see Figure). It is filled with liquid of dielectric constant $K_d = 2$ and connected to a fixed voltage V. The liquid slowly leaks out so that its level decreases with velocity v. Find the capacitance $C(t)$ and charge $Q(t)$ as a function of time t after the leak begins. Express your answer in terms of l, v and physical constants.

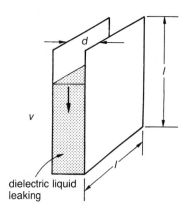

P261. A parallel plate capacitor has plate area A and holds charge Q. If the distance between the plates is x, find the total electrostatic energy stored in the capacitor. Hence show that the force between the plates is $F = -Q^2/2\epsilon_0 A$. A given capacitor has square plates of side $l = 10$ cm and is filled with material of dielectric constant $K_d = 3$. It is found that when the capacitor is uncharged and lying on its side it can support a mass of no more than 200 kg before collapsing. What is the maximum charge the capacitor can ever in principle hold? What happens to this maximum if K_d is halved?

P262. A parallel plate capacitor of area S and separation d (with $S \gg d^2$) is connected to a voltage source V through a switch. Calculate the charge Q on each plate, the electric field E between the plates, and the electrostatic energy U in each of the three cases below.
   (a) – The switch is closed and the system reaches a steady state.
   (b) – The switch is closed, the plates separation is increased to $2d$ and the system reaches a steady state.
   (c) – The switch is open, the plate separation is increased to $2d$; the switch is then closed and the system reaches steady state.
Express your answers in terms of S, d and V.

P263. Two conducting spheres, of radii $R_1 = 0.2$ m and $R_2 = 0.1$ m carry charges $q_1 = 6 \times 10^{-8}$ C, $q_2 = -2 \times 10^{-8}$ C and are placed at a distance $\gg R_1, R_2$ from each other. They are then connected by a conducting wire: what are their final charges?

P264. In the previous problem, find the total electrostatic energy of the two spheres before and after connection (neglect their interaction energy as they are very distant). Is it surprising that the two energies are not equal?

P265. A conducting sphere of radius $R_1 = 1$ m is charged by connecting it to a potential $V = 9 \times 10^3$ V. After it is fully charged it is disconnected. An uncharged conducting sphere of radius $R_2 = 2$ m is brought into electrical contact with the first sphere at large distance by means of a long wire and then disconnected. What are the charges on the two spheres now?

P266. Two spherical conducting shells have radii $R_1 = a, R_2 = 3a$ and equal charges q. What is the potential difference between them if they are:
   (a) – far apart,
   (b) – arranged with one concentrically inside the other?

P267. A point charge q is placed at the center of a perfectly conducting spherical shell of inner and outer radii $R, 2R$ (see Figure). Find the electric field and

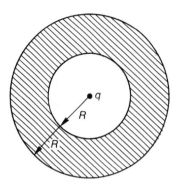

potential at radii $r_{out} > 2R, r_{in} < R$ and r_c with $R < r_c < 2R$. Repeat the calculation for the case where the shell is grounded (has zero potential).

P268. A plane parallel capacitor has square plates of side a and separation $d \ll a$ kept initially at a potential difference V. Material of dielectric constant $K_d = 2$ occupies half of the gap (see Figure). The material is now pulled slowly out of the capacitor. Find the capacitance $C(x)$ when the edge of the dielectric is a distance x from the center of the capacitor (see Figure). What current I flows in the circuit if the dielectric is removed at constant velocity u?

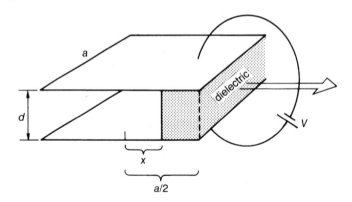

P269. Two thin concentric spherical shells of radii $R_A = R, R_B = 2R$ each carry uniformly distributed charge q. A third shell of radius $R_C = R$ and uniformly distributed charge $-2q$ is at a distance $\gg R$ from A, B. Calculate the electrostatic potential of each shell. If B and C are connected by a conducting wire, what will their potentials be once the system reaches a steady state?

P270. Electric fences are widely used in agriculture. If they are capable of giving a large cow a noticeable shock, how are small birds able to sit on them quite safely?

◼ ELECTRIC CURRENTS AND CIRCUITS

P271. A student uses a car battery (emf $\mathcal{E} = 12$ V) to power his electric razor. The battery supplies charge $Q = 0.5$ C each second. What electron current flows in the razor's motor and what power does the battery supply?

P272. A battery of emf $\mathcal{E} = 6$ V is connected to a resistance R. The current in the circuit is measured to be $I = 0.2$ A and the voltage drop across the battery is $V_0 = 5.8$ V. Find the internal resistance R_{in} of the battery.

P273. A battery of emf $\mathcal{E} = 10$ V and internal resistance $r = 1\,\Omega$ is connected to two resistors $R = 2\,\Omega$. Calculate the current drawn from the battery if the resistors R are connected:
  (a) – in series;
  (b) – in parallel.

P274. A copper pipe of length $l = 10$ m has inner and outer radii $r_1 = 0.9$ cm, $r_2 = 1$ cm. The resistivity of copper is $\rho_{Cu} = 1.75 \times 10^{-6}\,\Omega$ m. Find the resistance of the pipe.

P275. Find the resistance of a copper wire of length $l = 10$ cm if the wire has:
  (a) – cross-sectional area $A_1 = 3$ mm^2;
  (b) – cylindrical radius $r = 1$ cm. (The resistivity ρ of copper is given in the previous question.)

P276. Consider the circuit shown in the Figure. R_x is a variable resistor, and the internal resistance of the batteries is negligible. If the emfs $\mathcal{E}$ of the batteries are 6 V and $R_1 = R_2 = 2\,\Omega$, express the current I_2 in the resistor R_2 in terms of R_x. Is there a value of R_x for which this current vanishes?

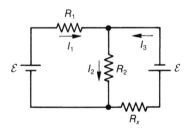

P277. Calculate the currents in the circuit in the Figure, where $\mathcal{E}_1 = 7$ V, $\mathcal{E}_2 = 3$ V, $R_1 = 4\,\Omega$, $R_2 = 5\,\Omega$, $R_3 = 8\,\Omega$, and the internal resistance of both batteries is negligible.

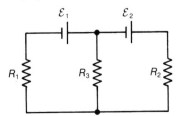

P278. Find the currents i_1, i_2 and i_3 at point A of the electrical circuit shown in the Figure.

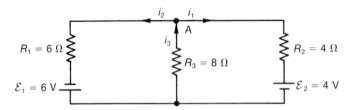

P279. A bulb and an emf source are to be connected in parallel across points A and B of the circuit shown in the Figure. What should the emf X be so that no current passes through the bulb?

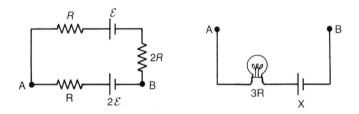

P280. An ammeter (of resistance R_A) and a voltmeter (of resistance R_V) are used to calibrate a resistor. If the resistor is connected as in Figure 1, the ammeter and voltmeter give readings I_1, V_1, while they read I_2, V_2 in the arrangement of Figure 2. The emf is the same in both cases. Express the resistance R in terms of the measured current and voltage and R_A, R_V in the two cases. Under what conditions is it correct to say that both methods give the resistance R as (measured voltage)/(measured current)?

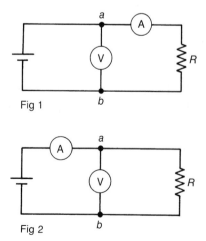

P281. An electric circuit consists of a power supply $\mathcal{E}$ and two equal resistors R in series (see Figure). A voltmeter of internal resistance r is used to measure the potential differences V_{cb}, V_{ba}. Find V_{cb}, V_{ba} in terms of $\mathcal{E}$, R and r.

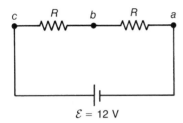

P282. Consider the three circuits (a, b, c) shown in the Figure. In which circuit is the dissipated electric power greatest? You may neglect the internal resistance of the power supply $\mathcal{E}$.

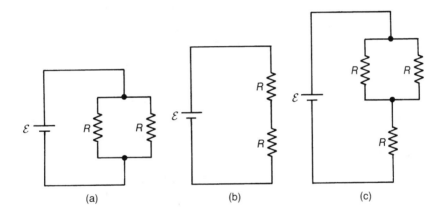

P283. An electric heater of resistance $R = 50\,\Omega$ is connected to a $V = 110$ V power supply for a time $t = 1$ h. How much energy is used?

P284. If the cost of 1 kWh of electrical energy is 30 cents, how much does it cost to use a 100 W lamp for 24 h?

P285. The starter motor of a car draws a current $I = 300$ A from the $V = 12$ V battery. What is the power consumption? If the car starts only after $t = 2$ min, how much energy is drawn from the battery?

P286. In the circuit shown in the Figure, the ammeter reading for the current is taken
    (a) – with both switches open;
    (b) – with both switches closed.

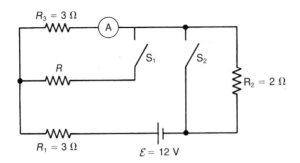

The readings are the same in the two cases. The power supply $\mathcal{E}$ has negligible internal resistance; using the values $R_1 = 3\,\Omega$, $R_2 = 2\,\Omega$, $R_3 = 3\,\Omega$ and $\mathcal{E} = 12$ V, find the resistance R.

P287. Father and son disagree about how to light their Christmas tree with 8 identical bulbs, using a battery of emf $\mathcal{E}$. The father wishes to connect the bulbs in series, while the son argues that the bulbs will be brighter if connected in parallel. Who is right?

P288. In a military exercise a field telephone is a distance $d = 5$ km from the command post. The wires have resistance $r = 6\,\Omega\,\mathrm{km}^{-1}$ and the telephone has resistance $R_T = 576\,\Omega$. Hoping to capture the line intact rather than simply destroying it, the "enemy" disables it by short-circuiting the pair of telephone wires with a metal rod of unknown resistance. To try to discover the problem, technicians measure the resistance R_c of the circuit twice: with the telephone connected they find $R_c = 120\,\Omega$, and with it disconnected they find $R_d = 150\,\Omega$. How far along the line from the command post is the problem? What is the resistance R_s of the metal rod causing the short?

P289. Two bulbs A, B of resistance $R, 2R$ are available to light a shared office and can be connected either in series or parallel. The clerk sitting under bulb A insists on connecting them so as to maximize the light from that bulb, while the other clerk argues that it is better to maximize the total light output. Can they agree on how to connect the bulbs? (Assume that the emitted light is proportional to the dissipated power.)

P290. Consider the circuit shown in the Figure. AB is a uniform wire of resistance $R_{AB} = 20\,\Omega$ and length 1 m. The point P is a moveable connection; when this

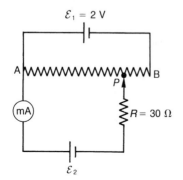

is placed 60 cm from A, the milliammeter registers zero current. Neglecting the internal resistances of the power supplies $\mathcal{E}_1, \mathcal{E}_2$, find $\mathcal{E}_2$ and the potential difference V_R across the resistor R.

The connection P is moved so that it is 50 cm from A. Find the current I in the milliammeter, and the new value of V_R.

P291. In the circuit shown in the Figure, an emf source $\mathcal{E} = 12$ V and internal resistance $r = 0.3\,\Omega$ is connected to two resistors $R_1 = 1.5\,\Omega$ and $R_2 = 1.2\,\Omega$. Two capacitors $C_1 = 0.05\,\mu$F and $C_2 = 0.02\,\mu$F are connected in parallel to the resistors, and the switch S is open. Calculate the current in the circuit and the charges Q_1, Q_2 on the capacitors once a steady state is reached. What values do these quantities take if the switch is closed and a new steady state is reached?

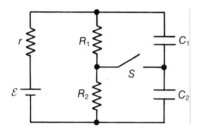

P292. In the circuit shown in the Figure, calculate the currents I_1, I_2 in R_1, R_2. What is the potential difference V_{AB}, and what are the charges on all three capacitors? ($\mathcal{E} = 10$ V, $R_1 = 1\,\Omega$, $R_2 = 4\,\Omega$, $C_1 = 1\,\mu$F, $C_2 = 5\,\mu$F.)

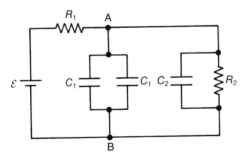

■ MAGNETIC FORCES AND FIELDS

P293. Two very long parallel wires are a distance $d = 1$ m apart and carry equal and opposite currents of strength $I = 1$ A. Find the magnetic field between the wires in their plane. An electron moves with velocity $v = c/2$ along the line exactly halfway between the two wires in their plane (i.e. parallel to one of the currents). Find the magnetic force on it. What happens if the velocity is reversed?

P294. Two very long parallel conducting wires carry currents $I_1 = 1$ A, $I_2 = 2$ A in opposite directions. They hang horizontally from pylons by pairs of insulating cables, each of length $a = 1$ m, and are a distance $d \ll a$ apart. The wires have mass m per unit length and the cables make angles θ to the vertical (see Figure). Find θ and the magnetic field at a point midway between the wires.

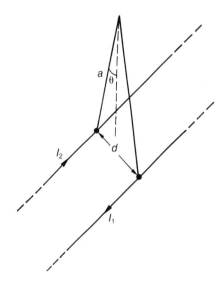

P295. A circular coil has $N = 10,000$ turns of wire arranged uniformly (see Figure). The wire carries current $I = 1$ A and the inner and outer radii of the coil are $a = 1$ m, $b = 2$ m. Describe the resultant magnetic field everywhere on the symmetry plane of the coil, and find the field strength at a distance $r = 1.5$ m from the center of the coil.

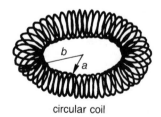

circular coil

P296. A slender solenoid of length $l = 1$ m is wound with two layers of wire. The inner layer has $N_1 = 1000$ turns and the outer one has $N_2 = 2000$ turns. Each carries the same current $I = 2$ A, but in opposite directions. What is the magnetic field inside the solenoid?

P297. A homeowner tries to set up a simple electric doorbell mechanism (see Figure). A permanent magnet of moment $\mu = 10^{-3}$ A.m is suspended by a wire that resists twisting. A solenoid of length $l = 10$ cm lies in the plane of the magnet, at an angle $\theta = 45°$ to its axis. Each loop of the solenoid has resistance $r = 10^{-3}$ Ω, and the solenoid is connected to a battery of emf $\mathcal{E} = 12$ V. A torque $T_s = 10^{-5}$ N.m is required to make the arm strike the bell: will the mechanism function? (Assume that the magnetic field of the solenoid at the permanent magnet is 0.01 of its value inside the solenoid.)

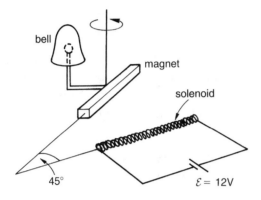

P298. Parallel loops of radii $r_0, 2r_0$ are a distance $d = 4r_0$ apart and carry currents I in opposite senses. Find the magnetic field B_P at the point P halfway between the loops as a function of I, r_0 and physical constants.

P299. A long wire carrying current $I = 10$ A lies in the plane of a rigid rectangular loop carrying current $I_1 = 1$ A (see Figure), parallel to its longer sides. The rectangle has sides $a = 0.2$ m, $b = 0.3$ m as shown, and the wire is $d = 0.25$ m from the loop. Find the magnitude and direction of the resultant force on the loop.

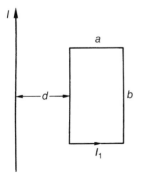

P300. The arrangement of the previous problem is used in the design of a magnetically levitated train. Many vertical loops (a rectangular coil) are fixed in

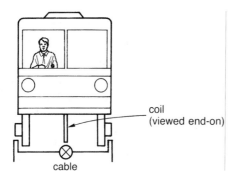

coil
(viewed end-on)

cable

each carriage directly above a cable fixed to the track bed (see Figure). The coil and carriage have the same length b. The carriage has weight per unit length w kg m^{-1}. How should the dimensions d, a be chosen so as to minimize the power requirements? If $d = 1$ cm, $w = 1000$ kg m^{-1} and the trackbed cable and coil each carry currents of 100 A, how many turns would the coil need?

P301. In the magnetically levitated train of the previous problem, three football players each weighing 100 kg take their seats in a particular 1 m section of a carriage. What happens to d?

P302. A long wire carrying a current $I = 1$ A is bent at its midpoint around one quarter of a circle of radius $r = 0.1$ m, the straight parts of the wire being perpendicular to each other (see Figure). Find the magnetic field at the point O.

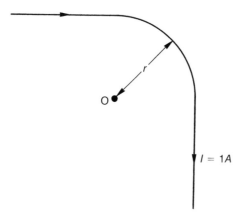

O

r

$I = 1A$

P303. A horizontal conducting rod of length L and mass m can slide on a vertical track (see Figure) and is in equilibrium at height L above a long horizontal wire when both the rod and wire carry current I, but in opposite directions.

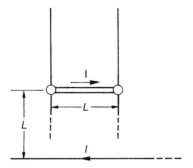

Find I in terms of m, L. If the current in the lower wire is suddenly doubled, what is the initial acceleration of the rod?

P304. A particle of charge q and mass m moves in the plane perpendicular to a uniform magnetic field B. Show that the particle moves in a circle, and find the angular frequency of the motion. What happens if the particle's velocity does not lie in the plane perpendicular to the field?

P305. A cyclotron is a device in which electrons gyrate in a uniform magnetic field B. In so doing they emit radio waves at the cyclotron frequency (see previous problem). The inventor of the cyclotron, Ernest O. Lawrence, was able to tell whether the apparatus was operating even when at home (and thus keep his graduate students up to the mark) by tuning a radio receiver to the appropriate wavelength and listening for the hum. Lawrence's original cyclotron had $B = 4.1 \times 10^{-4}$ T. What wavelength was his radio tuned to?

P306. Three long wires carry currents $I_1 = 8$ A (horizontally), $I_2 = I_1$ A (horizontally, but opposite to the first current), and $I_3 = I_1/2$ A (vertically downwards, perpendicular to the first two). Find the magnetic fields at the point P indicated in the Figure, with $a = 1$ m.

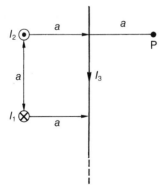

P307. A particle of charge q and mass m is accelerated from rest by a constant electric field E_0 acting over a length d (see Figure). It then encounters a region

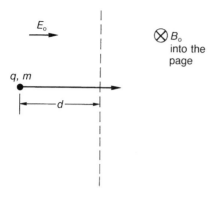

of constant magnetic field B_0 perpendicular to its velocity. Describe its subsequent motion. For what value of B_0 will the particle re-enter the region of constant electric field a distance d from the point at which it left?

P308. The arrangement of the previous problem can be used to measure the ratio q/m for unknown particles (the apparatus is called a *mass spectrometer*). Using the results of the previous problem, find q/m for a particle whose deflection $2R$ is measured to be D. If $E_0 = 10^5$ N/C, $d = 10$ cm, $B_0 = 0.1$ T and $D = 9.1$ cm, calculate q/m and compare it with the values for electrons and protons.

P309. Three types of particles are emitted by a certain radioactive sample. The particles are accelerated by a very large potential difference V and then enter a region of constant magnetic field B directed perpendicular to their motion. The radii of the particle orbits are in the ratio $R_1 : R_2 : R_3 = 1 : 2 : 3$ and their charges are equal. What can you infer about the particles' masses?

P310. A particle of mass m and charge q moves with constant velocity v along the negative x-axis, towards increasing x (see Figure). Between $x = 0$ and $x = b$ there is a region of uniform magnetic field B in the y-direction. Under what conditions will the particle reach the region $x > b$? If it does, at what angle to the x-axis will it enter this region?

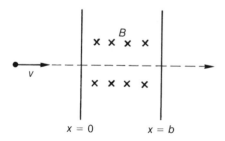

P311. A charged particle is injected with velocity v into a region containing electric and magnetic fields E, B, which are perpendicular to each other and also to

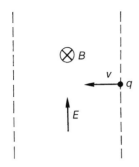

the particle's velocity (see Figure). E and B are adjusted so that the particle is undeflected. Find its velocity v in terms of E and B. How can this arrangement be used to select only particles of a particular speed from a beam with a range of speeds?

P312. A slender solenoid of length $L = 2$ m with $N = 10,000$ turns carries a current $I = 2$ A. Inside the solenoid, near the midpoint, there is a rectangular conducting loop $ABCD$ (see Figure) with plane parallel to the axis of the solenoid. The loop has $AB = 10$ cm, $BC = 6$ cm, and carries current $i = 1$ A. Find the resultant force and torque on the loop.

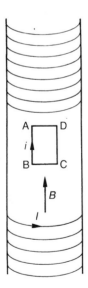

P313. A rectangular wire loop carries current I and is free to rotate about its long axis (length l) in a region of uniform magnetic field B. If its short axis has length w, show that when the loop plane makes an angle θ to the field (see Figure) the loop experiences a torque $BIlw\cos\theta$ about its axis. What

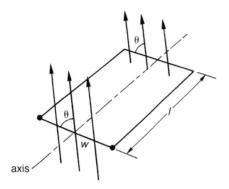

happens if the current I is reversed each time the loop is perpendicular to the field?

P314. A mass M with small electric charge q slides on a smooth inclined plane of angle θ to the horizontal. A magnetic field B is directed perpendicular to the section of the plane (see Figure). Calculate the acceleration of the mass when its velocity is u.

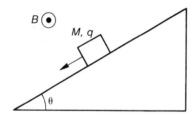

■ ELECTROMAGNETIC INDUCTION

P315. A rectangular wire loop with sides $l_1 = 0.5$ m, $l_2 = 1$ m is removed with constant velocity $v = 3$ m s^{-1} parallel to its longer sides from a region of constant magnetic field $B_0 = 1$ T perpendicular to its plane (see Figure). The loop's electrical resistance is $R = 1.5$ Ω. Find the current in the loop as a

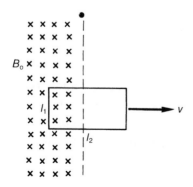

function of the distance x of its leading edge from the boundary of the field region.

P316. A plane loop of wire of area A is rotated about an axis lying in its plane, in a region of magnetic field B (see Figure). Show that a current flows alternately in the wire in one direction and then reverses symmetrically each time the loop is rotated. If the loop is rotated N times per second, show that the average induced emf in one half of the cycle is $2NAB$.

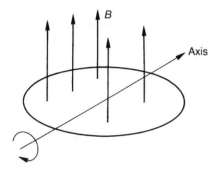

P317. An emf $\mathcal{E}_1$ is used to drive a current I_1 through a long solenoid of cross-sectional area A with n turns of wire per unit length and total resistance R_1. The emf alternates N times per second (see previous problem), and the solenoid is surrounded by a coil of m turns of wire per unit length. Show that the average emf induced in the coil over one half of the cycle is $\mathcal{E}_2 = 2NA\mu_0 nm\mathcal{E}_1/R_1$.

P318. The ends A, B of a conducting rod of length $l = 1$ m can slide freely while maintaining electrical contact with a rectangular conducting loop $KLMN$ (see Figure). A constant magnetic field $B_0 = 2$ T is directed perpendicular to the plane of the loop (into the page). Sides KM and LN have resistance $R_{KM} = 1\ \Omega$ and $R_{LN} = 2\ \Omega$ respectively, and the rest of the loop has negligible resistance. The rod AB is moved with constant velocity $v = 5\,\mathrm{m\,s}^{-1}$ towards LN. What force must be applied to maintain this motion?

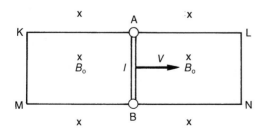

P319. A long conducting wire is bent at an angle of $60°$ and lies in a plane perpendicular to a uniform magnetic field $B_0 = 1$ T. A second very long conducting

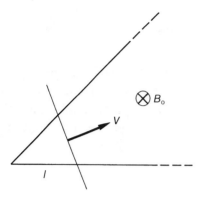

wire is pulled with velocity $v = 2$ m s^{-1} while lying on top of the bent wire so that the points of contact and the $60°$ vertex make an equilateral triangle (see Figure). At time $t = 0$ the triangle has side $l_0 = 0.5$ m. Both wires have uniform resistance per unit length $r = 0.1$ Ω m^{-1}. Assuming perfect contact between the two wires, express the induced emf in the triangle as a function of time in terms of B_0, v, l_0 and t. What is the value of this emf at $t = 5$ s? Find the current in the triangle at this time.

P320. An amusement park owner designs a new test-your-strength machine. Contestants propel a metal bob up a smooth vertical slide by means of a hammer

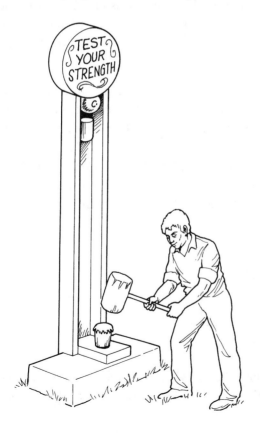

(see Figure). To measure the initial speed they give to the bob, the owner decides to use the induction effect in the Earth's magnetic field ($B = 10^{-4}$ T): the bob completes a circuit with the sides of the slide, and a voltmeter measures the induced emf. If the bob is $w = 10$ cm wide, and contestants typically manage to make the bob rise to heights $h = 10$ m, how sensitive must the voltmeter be?

P321. A plane conducting circular wire loop lies perpendicular to a uniform magnetic field B, and its area $S(t)$ is changed as $S(t) = S_0(1 - \alpha t)$ for $0 < t < 1/\alpha$ (S_0, α constant). The wire has resistance per unit length $\rho \, \Omega \, m^{-1}$. Find the current in the wire.

P322. A conducting loop of area $A = 1 \, m^2$ and $N = 200$ turns whose resistance is $R = 1.2 \, \Omega$ is situated in a region of constant external magnetic field $B = 0.6$ T parallel to its axis. The loop is removed from the field region in a time $t = 10^{-3}$ s. Calculate the total work done.

P323. A physicist works in a laboratory where the magnetic field is $B_1 = 2$ T. She wears a necklace enclosing area $A = 0.01 \, m^2$ of field and having a resistance $r = 0.01 \, \Omega$. Because of a power failure, the field decays to $B_2 = 1$ T in a time $t = 10^{-3}$ s. Estimate the current in her necklace and the total heat produced.

P324. To measure the field B between the poles of an electromagnet, a small test loop of area $A = 10^{-4} \, m^2$, resistance $R = 10 \, \Omega$ and $N = 20$ turns is pulled out of it. A galvanometer shows that a total charge $Q = 2 \times 10^{-6}$ C passed through the loop. What is B?

P325. A coil carries a current of $I = 10$ A. When the circuit is broken the current decays to zero in a time $\Delta t = 0.25$ s. The inductance of the coil is $L = 18$ Henry. What is the average induced emf?

P326. When a current in a certain coil varies at a rate of 50 A s^{-1} the induced emf is $V = 20$ volts. What is the inductance of the coil?

P327. A coil of $N = 100$ turns carries a current $I = 5$ A and creates a magnetic flux $\Phi = 10^{-5}$ T m^2. What is its inductance L?

P328. A rectangular loop of conducting wire has area A and N turns. It is free to rotate about an axis of symmetry. A constant magnetic field B is present and perpendicular to the axis. Find the induced emf as a function of time if the loop is rotated at angular velocity ω.

P329. A device for measuring wind speed has two conical cups attached to a horizontal rod of length $L = 0.5$ m (see Figure). The rod is attached to a vertical axle, which rotates a vertical conducting wire loop of area $A = 0.1 \, m^2$ and $N = 200$ turns. The Earth's magnetic field has horizontal component

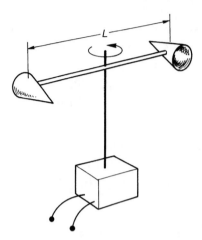

$B = 10^{-4}$ T at this point. Find the maximum voltage induced by a wind of speed $v = 100$ km/h, assuming that the cups rotate at exactly this speed.

CHAPTER THREE

MATTER AND WAVES

■ SUMMARY OF THEORY

I. Pressure

● A force F acting perpendicularly on an area A exerts (average) pressure

$$P = \frac{F}{A}. \tag{1}$$

● The *hydrostatic pressure* at depth h below the surface of a fluid of mass density ρ is

$$P = \rho g h \tag{2}$$

The hydrostatic pressure of the atmosphere is always close to $P_A = 10^5 \, \text{N m}^{-2}$ at sea level. P_A is called 1 atmosphere (1 atm).

● *Archimedes' principle* states that a body partly or wholly immersed in a fluid experiences a *buoyancy force* equal to the weight of the fluid it displaces. This force acts vertically upwards through the center of mass of the displaced fluid (the *center of flotation* or *buoyancy*).

2. Membranes and Surface Tension

Flexible enclosures such as balloons or tires exert a tension force resisting the pressure of their contents.

● A spherical enclosure of radius r made of material exerting tension t per unit length supports a pressure difference

$$P_i - P_e = \frac{2t}{r} \tag{3}$$

between its interior and exterior. This is known as Laplace's relation. For a cylindrical enclosure the corresponding relation is

$$P_i - P_o = \frac{t}{r}. \tag{4}$$

● The free surface of a liquid exerts a *surface tension* γ per unit length. A membrane made of such a liquid exerts tension per unit length $t = 2\gamma$. The force of the liquid surface on a container is $\gamma \cos \theta$ per unit length, where θ is the *contact angle*, which depends on the liquid and the material of the container.

3. Bernoulli's Theorem

An *incompressible* fluid is one whose density ρ may be taken as constant. Water is effectively incompressible under standard terrestrial conditions, and so is air if we do not consider sonic or supersonic motions.

● If the pressure in such a fluid is P at a point where the fluid velocity is v, Bernoulli's theorem states that

$$\frac{P}{\rho} + \frac{1}{2}\rho v^2 + gy = \text{constant} \tag{5}$$

along a streamline in the fluid. Here y is the vertical height above some reference level in the fluid. This can be thought of as an equation of conservation of mechanical energy for the fluid.

4. Ideal Gases

● A *mole* of a substance is an amount whose mass is a number of grams equal to the molecular mass divided by the mass of a hydrogen atom m_H (the *molar mass*). Thus the molar mass of carbon 12 is 12 g. *Note that the gram mole is not an SI unit.*

● At conditions far removed from those under which they liquefy or solidify, most common gases (air, hydrogen, oxygen, nitrogen, helium, etc.) can be regarded as *ideal* (or *perfect*): a fixed mass obeys the *ideal (or perfect) gas law*

$$PV = nRT \tag{6}$$

where P, V, and T are the pressure, volume, and absolute temperature T of the gas, and n is the number of moles of gas. R is the *universal gas constant*. We also use alternative forms of this relation, such as

$$P = \frac{k}{\mu m_H} \rho T, \tag{7}$$

where ρ is the mass density of the gas, k is *Boltzmann's constant* and μ is the *mean molecular mass*, i.e. the mass of one molecule of the gas in units of the mass m_H of a hydrogen atom. This is consistent with the earlier form using R if it is remembered that the gram mole is not an SI unit. It is also sometimes convenient to use the form $PV = nRT$ with P in atm and V in liters. The appropriate value of R can be found in the table of constants.

The absolute temperature T (measured in K) and the temperature t (measured in °C) are related by $T = t + 273$.

5. Heat and Thermodynamics

● The *coefficient of linear expansion* α is the fractional length by which a solid expands when heated through 1°C. The *coefficient of volume expansion* γ is the fractional volume increase when the solid is heated through 1°C.

● The *specific heat* of a substance is the amount of heat required to raise the temperature of unit mass of it by 1°C.

● The *mechanical equivalent of heat* is approximately 4184 J/kcal, where 1 kcal (kilocalorie) is the amount of heat required to raise the temperature of 1 kg of water through 1°C.

● The *first law of thermodynamics* expresses the conservation of heat and mechanical energy in the form

$$\Delta Q = \Delta U + \Delta W. \tag{8}$$

Here ΔQ is the heat energy flowing into the system, ΔU is the increase in internal energy of the system, and ΔW is the work done by the system on its surroundings. For example, a gas of pressure P whose volume increases by ΔV performs work $\Delta W = P\Delta V$.

In an *adiabatic* process no heat is transferred to or from the system, so $\Delta U + \Delta W = 0$.

● The *second law of thermodynamics* states that *heat flows from hotter to colder bodies*; reverse flows can be arranged, but only at the cost of supplying energy to the system. When a system at absolute temperature T absorbs heat energy ΔQ at equilibrium (i.e. slowly), its *entropy* S changes by an amount

$$\Delta S = \frac{\Delta Q}{T}. \tag{9}$$

● If a body of mass m and specific heat C per unit mass is heated from T_1 to T_2, the total entropy change is

$$\Delta S = mC \ln \left(\frac{T_2}{T_1} \right). \tag{10}$$

● Clearly the entropy remains constant in an adiabatic change. The second law of thermodynamics can be restated in the form *the entropy of a closed system can never decrease*. The entropy of an ideal gas of pressure P occupying volume V remains constant if the quantity PV^γ is constant, where γ is the ratio of specific heats at constant pressure and constant volume. For an ideal monatomic gas $\gamma = 5/3$, and the full expression for the entropy is

$$S = \frac{3k}{2\mu m_H} \ln T + \frac{k}{\mu m_H} \ln V. \tag{11}$$

Using the ideal gas law to replace T by P, V this indeed shows that $PV^{5/3} = $ constant if S is constant. The internal energy of an ideal monatomic gas is

$$U = \frac{3k}{2\mu m_H} nRT. \tag{12}$$

For a diatomic gas (e.g. O_2) $\gamma = 7/5$.

6. Kinetic Theory of Gases

● Kinetic theory treats gases as composed of discrete particles or molecules in random motion.

The ideal gas law can be derived from the assumption that collisions of the gas particles are perfectly elastic. The average kinetic energy of the particles is $3kT/2$, where k is Boltzmann's constant, so their average (root-mean-square) speed is

$$v_{\text{rms}} = \left(\frac{3kT}{\mu m_H} \right)^{1/2}. \tag{13}$$

7. Light

● *Refraction* of light is governed by two laws:
  1. – At a boundary between two media, the incident and refracted rays and the normal to the boundary all lie in the same plane.
  2. – The angles of incidence and refraction θ_1, θ_2 are related by

$$n_1 \sin \theta_1 = n_2 \sin \theta_2, \tag{14}$$

(Snell's law), where n_1, n_2 are the refractive indices of the media containing the two rays, and the angles are measured from the normal to the interface.

● For spherical *mirrors* of curvature radius R we adopt the following conventions: the focal length $f = -R/2$, where $R < 0$ for a concave mirror and $R > 0$ for a convex mirror. The object distance is s from the front of the mirror, and the image distance is s' behind the mirror. These quantities are related by the *mirror equation*

$$\frac{1}{s} - \frac{1}{s'} = \frac{1}{f}. \tag{15}$$

The image is *virtual* or *imaginary* if $s' > 0$ and *real* if $s' < 0$. The magnification $m = s'/s$ is positive for an upright image and negative for an inverted image.

● For *thin lenses*, we adopt the convention that the focal length $f > 0$ for converging lenses and $f < 0$ for diverging lenses. The object distance s is always positive and the image distance s' is positive when it is on the opposite side of the lens. These quantities are related by the *thin lens equation*

$$\frac{1}{s} + \frac{1}{s'} = \frac{1}{f}. \tag{16}$$

A virtual image has $s' < 0$. The magnification $m = -s'/s$ is positive for an upright image and negative for an inverted image.

● The focal length f of a thin lens made of material of refractive index n is given by the *lensmaker's equation*

$$\frac{1}{f} = (n - 1)\left(\frac{1}{R_1} + \frac{1}{R_2}\right), \tag{17}$$

where R_1, R_2 are the curvature radii of its two faces, counted positive if they are convex and negative if concave.

● The quantity $P = 1/f$ is called the *power* of a lens, and is measured in $m^{-1} =$ diopters.

A mirror or lens is denoted $f/4$ or $f/8$, etc. if its diameter is 1/4 or 1/8 of its focal length f.

● A *wave* disturbance (e.g. light, sound) propagating in the x-direction can be represented as

$$\psi(x, t) = A \sin\left[2\pi\nu t - \frac{2\pi}{\lambda}x\right]. \tag{18}$$

Here A is the *amplitude*, ν the *frequency* [measured in Hertz (Hz) = cycles s^{-1}] and λ the *wavelength*. The combination in square brackets is the *phase* $\phi(x, t)$. The *phase velocity* $v_\phi = \lambda\nu$. One sometimes also uses the *angular frequency* $\omega = 2\pi\nu$, which is measured in radians s^{-1}.

● A wave emitter in motion exhibits the *Doppler effect*: the frequency of the received waves is raised (lowered) if the motion is towards (away from) the observer. For light waves the frequency change is

$$\frac{\Delta\nu}{\nu} = -\frac{v}{c} \tag{19}$$

where c is the phase velocity of the wave and v is the velocity along the line joining the observer to the emitter: $v > 0$ implies motion *away* from the observer. The corresponding wavelength change is

$$\frac{\Delta\lambda}{\lambda} = \frac{v}{c}. \tag{20}$$

For sound waves the source velocity is added to the phase velocity, so a stationary observer hears the frequency

$$\nu = \nu_0 \frac{v_s}{v_s + v}, \tag{21}$$

or wavelength

$$\lambda = \lambda_0 \frac{v_s + v}{v_s}. \tag{22}$$

Here v_s is the velocity of sound, v is the velocity of the source away from the observer, and the suffix 0 refers to the frequency and wavelength for a source at rest.

● *Coherent* waves have the same frequency and a fixed phase difference. *Interference* occurs when two or more coherent waves interact. If the waves have the same phase where they are combined, we have *constructive* interference (e.g. greater light intensity); if they have phases that differ by π radians = 180°, this is *destructive* interference (reduced light intensity).

When parallel light rays of wavelength λ are normally incident on two slits separated by distance d, *interference* fringes are observed. Constructive interference occurs at angles θ_n to the original ray direction, where

$$d \sin\theta_n = n\lambda, \quad n = 0, 1, 2, \ldots \tag{23}$$

This is also true for a *diffraction grating* with spacing d.

Diffraction from a single slit of width D produces destructive interference at angles θ_m to the original direction, where

$$D \sin \theta_m = \pm m\lambda, \ m = 1, 2, 3, \ldots \qquad (24)$$

8. Atomic Physics

- The energy of a photon of frequency ν is $E = h\nu$, where h is Planck's constant. The momentum of the photon is $p = E/c = h\nu/c = h/\lambda$.

- The *de Broglie* wavelength of a body of momentum p is $\lambda_B = h/p$.

- The *uncertainty principle* states that the uncertainties $\Delta x, \Delta p$ in position and momentum obey the inequality

$$\Delta x \Delta p \gtrsim \hbar, \qquad (25)$$

where $\hbar = h/(2\pi)$.

- In the *photoelectric effect*, incident light of wavelength λ releases a photoelectron of energy

$$E_k = \frac{hc}{\lambda} - B, \qquad (26)$$

where B is a constant called the *work function* of the medium surface.

- Light scattered through an angle θ by free electrons of mass m_e has its wavelength λ changed to λ', where

$$\lambda' = \lambda + \lambda_c(1 - \cos \theta). \qquad (27)$$

Here $\lambda_c = h/m_e c = 0.024$ Å is the *Compton wavelength* of the electron, and this is called *Compton scattering*. The Angstrom unit (Å) is defined by $1\text{Å} = 10^{-10}$ m.

- The energy levels of the *Bohr model* of the hydrogen atom are

$$E_n = -\frac{E_0}{n^2}, \qquad (28)$$

where $E_0 = 13.6$ eV is the *Rydberg* and n is the *principal quantum number*, which takes integer values. When the electron jumps between these levels, the energy of the emitted or absorbed photon is given by the difference $E_n - E_m$. The transitions down to $n = 1$ give *spectral lines* called the *Lyman* series, and those to $n = 2$ the *Balmer* series. The lines appear in absorption if there is a cooler transparent medium in front of a hotter one. In the limit $n = \infty$ the electron is no longer bound to the atom, which is therefore *ionized*. The *ionization potential* is the energy required to bring this about, which is $I_n = E_0/n^2$ for ionization from the nth bound level.

In *radioactive decay* the number of radioactive nuclei decreases in time according to

$$N(t) = N_0 e^{-\lambda t}, \qquad (29)$$

where λ is the *decay constant*, characteristic of the nucleus, and $e = 2.718$ is the base of natural logarithms. The *half-life* $t_{1/2}$ is the time in which one-half of a large sample of the nuclei will decay. It is related to λ by $\lambda t_{1/2} = 0.693$. The *activity* of the nucleus is defined by

$$A = -\frac{\Delta N}{\Delta t}, \qquad (30)$$

where ΔN is the change in the number of nuclei in time interval Δt: one can show that $A = \lambda N(t)$.

Nuclei of the same charge number Z but different mass number A are called *isotopes*.

In *beta decay* a neutron disintegrates into a proton, an electron and an antineutrino. This increases Z by one but leaves A unchanged.

9. Relativity

The theory of relativity is based on the postulate that the *velocity of light in free space is the same for all observers*. As a consequence, observers moving relative to each other with velocity v assign different values to various physical quantities. The relations between them involve the quantity

$$\gamma(v) = \left(1 - \frac{v^2}{c^2}\right)^{-1/2}. \qquad (31)$$

Time dilation. A time interval t_0 on a clock at rest with respect to an observer is seen as having the value t when in motion, where

$$t = \gamma t_0. \qquad (32)$$

t_0 is the *proper time*.

Length contraction. An object of length l_0 when at rest with respect to an observer ($l_0 =$ the *proper length*) appears shortened to length l when in motion, where

$$l = \frac{l_0}{\gamma}. \qquad (33)$$

● *Simultaneity.* Events occurring at different points but at the same instant for one observer do not in general appear simultaneous for another observer.

● *Relativistic velocity addition formula.* If an object is seen by observer 1 to move at velocity v_1, and observer 1 is seen by a second observer (2) to move at velocity v_2 in the same direction, then observer 2 sees the object moving with velocity

$$V = \frac{v_1 + v_2}{1 + v_1 v_2 / c^2}.$$
(34)

Thus V can never exceed c; *no object can be accelerated to speeds $> c$.*

● The *energy* of a body of rest-mass m moving at speed v is

$$E = \gamma m c^2.$$
(35)

It therefore has *rest-mass energy $E_0 = mc^2$* when $v = 0$. The *momentum* of the body is

$$p = \gamma m v.$$
(36)

These two quantities are related by

$$E^2 = p^2 c^2 + m^2 c^4.$$
(37)

■ LIQUIDS AND GASES

P330. Oil is added to the right-hand arm of a U-tube containing water. The oil floats above the water to a height of $h = 10$ cm. The top of the oil + water column is a height $d = 2$ cm above the top of the water column in the other arm (see Figure). Calculate the oil density ρ_0. Fluid of density ρ_x is added to the water column in the left arm to a height $l = h/2$. If the fluid levels in the two arms are now equal, calculate ρ_x.

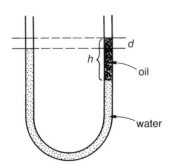

P331. A hydraulic press contains oil of density $\rho_0 = 800$ kg m^{-3}, and the areas of the large and small cylinders are $A_l = 0.5$ m^2, $A_s = 10^{-4}$ m^2. The mass of the large piston is $M_l = 51$ kg, while the small piston has an unknown mass m. If an additional mass $M = 510$ kg is placed on the large piston, the press is in balance with the small piston a height $h = 1$ m above the large one (see Figure). Find the mass m.

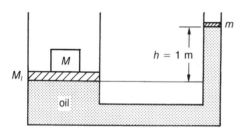

P332. How could you decide if a wedding ring is made of pure gold using sensitive scales, a liquid volume measure, a length of thread, and a sample of pure gold?

P333. A woman of mass $M = 60$ kg has height $h = 1.6$ m and shoulder width $w = 45$ cm. She wears shoes of length $l = 25$ cm and average breadth $b = 7$ cm. Approximating the relevant areas as rectangles, what average pressure does she exert
(a) – on the ground when standing,
(b) – on a bed when lying flat?
Why is it uncomfortable to lie on a hard floor? What pressure does the woman exert if she puts her weight on stiletto heels of total area $A = 2$ cm^2?

P334. The tires on a racing bicycle are inflated to a pressure $P = 7$ atm. Does the pressure gauge on the pump read 7 atm? The combined mass of the bicycle and rider is $m = 70$ kg. What is the total tire area in contact with the road?

P335. Two cylinders of cross-sectional area $A = 10$ m^2 are fitted smoothly together as shown in the Figure, and then evacuated. Masses M are hung from cables attached to each of the cylinders. How large can the masses M be made before the cylinders are pulled apart?

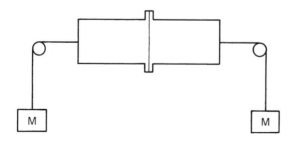

P336. A payload $m = 200\,\text{kg}$ is held stationary by a balloon at a certain height above the ground. The volume of the balloon is $V_b = 1000\,\text{m}^3$, and is far larger than that of the payload. Express the gas density ρ_b inside the balloon in terms of the air density ρ_a at this height.

P337. Early airships were filled with hydrogen rather than with helium, sometimes with tragic consequences (e.g. the destruction by fire of the German airship *Hindenburg* in 1937). One sometimes reads that the reason for using hydrogen was that, since the density ρ_{He} of helium is twice that of hydrogen (ρ_{H}) under the same conditions, twice the volume of helium would have been needed to lift the same payload. Is this correct? ($\rho_{\text{H}} = 0.09\,\text{kg m}^{-3}$, air density $\rho_a = 1.3\,\text{kg m}^{-3}$.)

P338. A ball of uniform density 2/3 of that of water falls vertically into a pond from a height $h = 10$ m above its surface. How deep below the surface can the ball sink before buoyancy forces push it back? (Neglect the water drag on the motion of the ball.)

P339. A yacht is at rest on a small lake. What happens to the water level if the yachtsman throws overboard (a) a buoy, and (b) an anchor?

P340. A plastic cube of density $\rho = 800\,\text{kg m}^{-3}$ and side $a = 5\,\text{cm}$ is floated in a cylindrical water container of surface area $A = 100\,\text{cm}^2$. Find the resulting increase h of the water height. A mass m is placed on the cube and just submerges it. Find m.

P341. A wooden cube of side $a = 0.1\,\text{m}$ is just submerged in water when pressed down with a force $F = 3.43\,\text{N}$. Calculate the density ρ of the wood. What depth of the cube is submerged if it floats freely?

P342. A cube of side a is made of material of density $\rho = 3\rho_w/4$, where ρ_w is the density of water. It is placed in a container with a square cross-section whose side is $a + c$, where $c \ll a$, and whose height exceeds a (see Figure). Find the

minimum volume V of water that must be poured into the container to float the cube. Can V be made arbitrarily small by reducing c?

P343. A solid cube of side $a = 0.1$ m hangs from a dynamometer (a spring measuring force), and is submerged inside a container of liquid. The container holds water, with above it a layer $d = 0.2$ m of oil of density $\rho_o = 500\,\mathrm{kg\,m^{-3}}$. In equilibrium the base of the cube is a distance $h = 0.02$ m below the water level (see Figure), so that its upper face is below the surface of the oil. The dynamometer reading is $W_D = 0.49$ N. Calculate the mass M of the cube and the hydrostatic pressure P at the base of the cube.

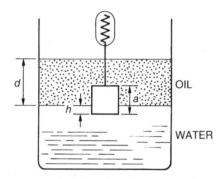

P344. An iceberg has the shape of a cube and floats in seawater with $h = 2.5$ m protruding above the surface. The densities of ice, seawater, and fresh water are $\rho_i = 900$ kg m^{-3}, $\rho_s = 1300$ kg m^{-3} and $\rho_f = 1000$ kg m^{-3} respectively. Find
   (a) – the submerged depth x_s of the iceberg in the sea,
   (b) – the submerged depth x_f in fresh water.
What fraction of the iceberg would be above the surface in the second case?

P345. A certain liquid has density ρ, and surface tension γ and contact angle θ when in contact with glass and air. Find the height h of the liquid in a glass tube of cylindrical radius r immersed in this liquid.

P346. Can capillary action account for sap rising in trees? (Assume surface tension of sap is $\gamma = 0.07\,\mathrm{N\,m^{-1}}$, contact angle $\theta = 0$, sap density $\rho = 10^3\,\mathrm{kg\,m^{-3}}$, tree capillary radius $= 10^{-2}$ mm.)

P347. A glass tube has a removable cap at one end, which tends to fall off when the tube is inverted. The cap is made of material of density $\rho = 700$ kg m^{-3} and is $d = 2$ mm thick. For what tube radii r will wetting the end of the tube keep the cap on when it is inverted? (Assume surface tension of water $\gamma = 0.07\,\mathrm{N\,m^{-1}}$ and contact angles $\theta = 0$ where appropriate.)

P348. The pressures inside and outside a spherical membrane of radius r are P_i, P_o, with $P_i > P_o$. Show that the material of the membrane must exert total tension per unit length t, where

$$P_i - P_o = \frac{2t}{r}.$$

If the material is a liquid whose *surface* tension is γ, show that

$$P_i - P_o = \frac{4\gamma}{r}.$$

P349. Repeat the last question for the case of a cylindrical tube of radius r. Why do boiling frankfurters tend to split lengthways rather than around their cross-sections?

P350. A tire on a racing bicycle is inflated to a pressure $P_i = 7$ atm. The radius of the tire is $r = 1.5$ cm. Find the tension in the walls.

P351. What is the radius r of the smallest droplet that can form from water of surface tension $\gamma = 0.07$ N m^{-1} and vapor pressure $P_v = 2300$ N m^{-2}?

P352. A spherical balloon has interior pressure P_1 and radius r_1, and is in equilibrium inside an enclosure with pressure $P_o = 8P_1/9$. The enclosure is gradually evacuated. Assuming that the temperature is fixed and the tension t per unit length of the balloon material remains constant, show that the balloon radius never exceeds $3r_1$.

P353. The air sacs in the lungs (alveoli) can be approximated as small spherical membranes of radius r containing air at atmospheric pressure P_0. The pressure P_c in the chest cavity (pleural pressure) increases when the person breathes out. Simultaneously, muscle contraction decreases r. These changes are reversed as the person breathes in. Show that the membrane tension per unit length t must decrease as the person exhales and increase as he inhales.

P354. Two identical small balloons are inflated, one much more than the other. They are then connected by a pipe which is closed by a valve between them. The whole apparatus is placed in an evacuated enclosure. What happens when the valve is opened?

(You may assume that the surface tension of the balloon material is independent of the balloon's size except when the balloon is smaller than a certain spherical radius r_{min}, below which the surface tension decreases.)

P355. A container is filled with water to a depth $H = 2.5$ m. The container is tightly sealed and above the water is air at pressure $P_1 = 1.34 \times 10^5$ N m^{-2} (see Figure). A small hole is drilled at a height $h = 1$ m above the bottom of the container. What is the speed of the resulting jet of water? Compare

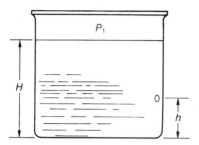

your answer with the case of a container open at the top, but otherwise identical.

P356. When a doctor measures a patient's blood pressure, the cuff is always placed around the arm, rather than the ankle or other part of the body. Why?

P357. A homeowner wishes to drain her swimming pool by siphoning the water, whose depth is h, into a nearby gully a distance H below it, where H is much larger than h (see Figure). She uses a pipe of cross-sectional area a, and the pool water has surface area A. How long does it take to empty the pool if $h = 2$ m, $H = 20$ m, $A = 50$ m^2, $a = 5$ cm^2?

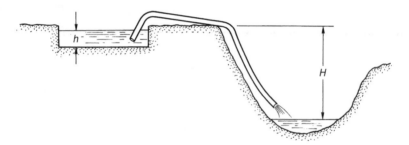

P358. In the siphon arrangement of the last question, the pipe develops a leak at a point above the water surface. What happens to the water flow? If there is no leak, what is the effect of having air trapped in the pipe?

P359. Water is pumped at a constant rate $r = 6$ m^3 min^{-1} through a pipe. Near the pump the pipe diameter is $d_1 = 0.2$ m, but this widens to a diameter

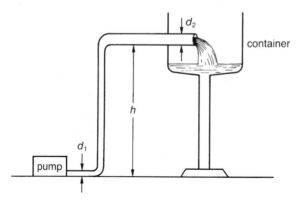

$d_2 = 0.4$ m in a horizontal section at a height $h = 20$ m above the pump (see Figure). This section discharges into a container open to the atmosphere. At what velocity does water leave the pipe?

P360. In the last problem, what is the water pressure near the pump?

P361. A wide container is filled with water up to a depth H. A small hole is drilled in the container at a distance h below the water level, and a jet of water emerges from it. How far from the container does the jet hit the ground?

P362. A Venturi tube (see Figure) is used to measure the water speed v in a pipe by comparing the pressures in the wide and narrow sections (cross-sectional areas $A, A' = A/4$). Find v if the difference in mercury levels is $h = 25$ mm. (The density of mercury is $\rho_{Hg} = 13,600$ kg m^{-3}.)

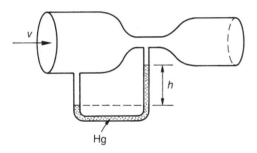

P363. The window and door of a room are both open. The door opens inwards: why does it tend to slam shut if only slightly ajar?

P364. Air of density $\rho = 1$ kg m^{-3} flows smoothly and horizontally over the airfoil shape shown in the Figure. The streamline path of air flowing above the airflow is m times longer than that of the air flowing below it, which has speed v. Show that the airfoil experiences an upward force

$$L \approx \frac{1}{2}(m^2 - 1)\rho v^2$$

per unit area. Assume that both streamlines pass through A and B.

An airplane of mass $M = 500$ kg has a total wing area $A = 30$ m^2, and the airfoil design is such that $m = 1.1$. Estimate the airplane's minimum takeoff speed at sea level ($\rho = 1$ kg m^{-3}). How does this change in high-altitude airports?

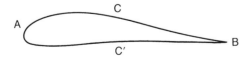

P365. At high altitude the airplane of the last question can achieve a maximum airspeed of $v_{max} = 70\,\mathrm{m\,s^{-1}}$. The air density ρ decreases with height z as $\rho = 1 \times 10^{-z/H}\,\mathrm{kg\,m^{-3}}$, where $H = 23{,}000$ m. What is the maximum height that the airplane can in principle achieve?

P366. In light of P364 can you suggest why the early airplanes (e.g. the Wright brothers') were all biplanes?

P367. Two species of bird are very similar in every respect except that every dimension of one species is on average l times the corresponding dimension of the other. How are their respective takeoff speeds for flight related?

P368. A *hydrofoil* boat uses submerged fins with airfoil-type cross-sections to lift the boat largely clear of the water and allow much higher speeds. Find the condition for this to be achieved at water speed u and total hydrofoil area A_h, if the water streamline path over the upper surface of the latter is m times longer than over the lower surface and the boat has mass M. Show that A_f can be much smaller than the wing area required for takeoff of an airplane of the same mass, even with slower speeds u (compare P364).

P369. When a yacht sails into the wind its sails adopt a curved shape as viewed from above (see Figure). At a suitable angle to the wind direction the air on the concave side of the sails moves much more slowly than that on the convex side. If the average speed of the latter is w, the sails have total area A, and the yacht steers at angle θ to the wind direction (see Figure), show that the yacht experiences total wind force

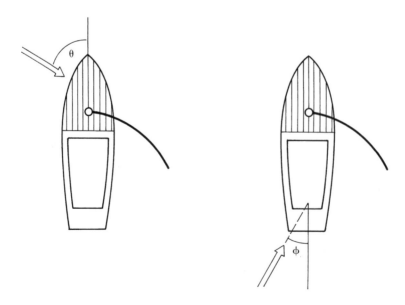

$$F_1 \approx \frac{A}{2} \rho_a w^2 \sin \theta$$

in the forward direction, where ρ_a is the air density.

The same yacht now sails with the wind more nearly behind it (see Figure: at an angle $\phi < 90°$ from directly astern). If the wind has velocity w and the boat's forward speed is much less than this, find the maximum forward force F_2 on the yacht and compare it with F_1 in the case $\theta = \phi = 45°$.

P370. The yacht of the previous question has submerged frontal cross-sectional area A_f (see Figure). If the water density is ρ and the yacht moves at speed v, show that it has to supply momentum $\approx A_f \rho v^2$ per unit time to the water, and thus estimate the drag force on it. Estimate the boat's speed v_1, v_2 in terms of $w, A, A_f, \rho_a, \rho, \theta$ and ϕ in the cases where it (1) sails into the wind, and (2) has the wind behind it. Evaluate v_1, v_2 for $w = 30$ km/h, $A = 20$ m^2, $A_f = 0.3$ m^2, $\theta = \phi = 45°$, using $\rho_a = 1$ kg m^{-3}, $\rho = 10^3$ kg m^{-3}.

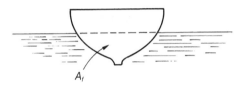

P371. The yacht considered in the last two problems has mass M, and the submerged depth is approximately constant along its length l. The sails are triangular and the mast has height l also (see Figure). Show that $A_f \approx Mg/\rho l$, and hence that the yacht should be designed to maximize the quantity l^3/M to achieve high speeds.

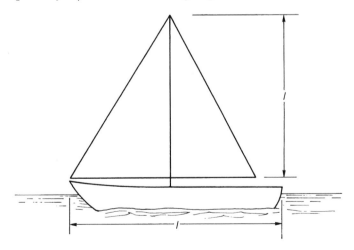

P372. A yacht, as considered in the previous three problems, resists sideways motions by means of its keel, which gives the boat total side-on cross-sectional area A_s. Show that the boat should be designed so that $A_s \gg A_f$. What is the usual way of achieving this?

P373. A certain Grand Prix racing car has mass $m = 1000$ kg, and the coefficient of sliding friction between its tires and the road is $\mu = 0.5$. What is the maximum speed at which it can take a level bend of radius of curvature $r = 100$ m?

A very efficient wing of area $A = 2$ m^2 is now fitted to the car, so that the air passing above the wing moves much more slowly than the car's speed v, while that passing below moves at v. What is the new maximum speed around the bend? (Air density $\rho = 1$ kg m^{-3}.)

P374. Is the wing of the last question more of an advantage on slow, tight corners or fast, relatively gentle ones?

P375. An ideal gas at temperature $t_1 = 16°C$ is heated until its pressure and volume are doubled: what is its final temperature?

P376. A closed container of volume $V_1 = 12$ liters holds a mass $m_1 = 0.858$ kg of oxygen. It is known that the mass of a liter of oxygen at atmospheric pressure is $m_2 = 0.0015$ kg at the same temperature. What is the pressure in the container?

P377. A cylindrical container is enclosed by a piston of mass $m = 21$ kg and holds a mass $m_H = 0.17$ g of molecular hydrogen. The volume of hydrogen is $V_H = 1400$ cm^3 and the height of the piston is $h = 40$ cm (see Figure). Find the atmospheric pressure P_A outside the container if the absolute temperature is $T = 300$ K.

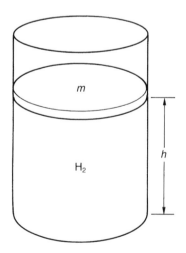

P378. A glass pipe of constant cross-sectional area $A = 10^{-4}$ m^2 and length $l = 1.14$ m is sealed at one end and closed by a cork at the other. Inside

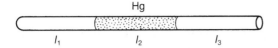

Hg

l_1 l_2 l_3

the pipe there is a mercury column of length $l_2 = 0.3$ m. When the cork is removed and the pipe held horizontally in the atmosphere, the air columns on each side of the mercury have equal lengths $l_1 = l_3 = 0.42$ m (see Figure). The pipe is now held vertically with the open end upwards. Find the length l_1' of the air column at its sealed end. What would be the length l_1'' of this column if instead the pipe had been corked in the horizontal position before being turned vertical? Assume that the temperature remains constant throughout. The density of mercury is $\rho_{Hg} = 13{,}600\ \text{kg}\,\text{m}^{-2}$.

P379. A glass bulb of radius $R = 1.5$ cm is attached to a glass tube of cross-sectional area $A = 0.2$ cm^2. A mercury drop of length $l_H = 6$ cm seals the air in the bulb and a length l_A of the tube (see Figure). When the temperature is $t = 10°C$ and the tube is horizontal, we have $l_A = 17$ cm; when the temperature is $t' = 20°C$ and the tube is vertical with the bulb at the bottom, we have $l_A' = 13.3$ cm. Find the atmospheric pressure P_A, given that the density of mercury is $\rho_{Hg} = 13{,}600$ kg m^{-3}. (Assume constant temperature.)

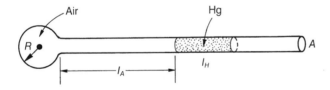

Air   Hg

R A

l_A l_H

P380. A narrow glass tube of length $l = 0.5$ m is sealed at one end. The open end is lowered vertically into a bath of mercury, which enters the tube and traps some air in the upper end. When the sealed end of the tube is $h_1 = 0.05$ m above the mercury level in the bath the mercury level in the tube is $h_2 = 0.15$ m below this level (see Figure 1). The tube is now raised so that the sealed end is $h_1' = 0.45$ m above the mercury level in the bath; the

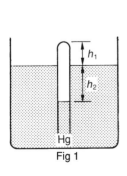

h_1

h_2

Hg

Fig 1

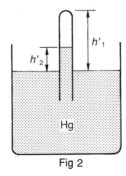

h'_1

h'_2

Hg

Fig 2

level in the tube is now $h_2' = 0.15$ m *above* this level (see Figure 2). Find the atmospheric pressure P_A. At what height h must the sealed end of the tube be placed so that the mercury in the tube is level with that in the bath? (Density of mercury $\rho_{Hg} = 13,600$ kg m^{-3}; assume constant temperature.)

P381. A solid cylinder of radius $R = 0.5$ m and height $H = 1$ m is drilled at one end to make a concentric cylindrical cavity of radius $r = R/2$ and depth $h = H/2$. The cylinder is placed in a large mercury bath with the drilled end lowest, and floats with its upper face exactly at the level of the mercury (see Figure). The atmospheric pressure is $P_A = 0.987 \times 10^5$ N. Calculate the pressure P_1 of the air trapped in the cavity, the height y of the mercury in the cavity above the cylinder's base, and the density ρ of the cylinder material. (Density of mercury $= 13,600$ kg m^{-3}.)

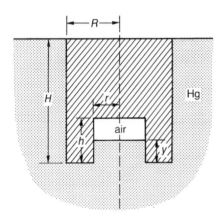

P382. Two containers of volumes $V_1 = 2V$, $V_2 = V$ are connected by a narrow pipe with a faucet (see Figure). With the faucet closed V_1, V_2 contain $n, 2n$ moles of a certain ideal gas respectively. The faucet is opened and the system allowed to stabilize at constant temperature. Find the number of moles in each container in terms of n.

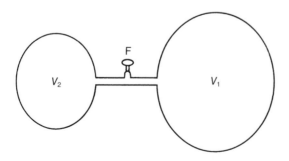

P383. Two containers of volumes $V_1 = 5$ liters and $V_2 = 3$ liters are connected by a narrow pipe with a faucet. The larger container has a valve, which releases gas if its pressure P_1 exceeds a value $P_{crit} = 3$ atm. The absolute temperature is $T = 275$ K, and with the faucet closed the containers hold ideal gas at pressures $P_1 = 2$ atm, $P_2 = 4$ atm. What is the total number of moles in the two containers? The faucet is now opened: does gas leak from the valve? If the system is heated to $T' = 400$ K how many moles of gas will remain in the containers?

■ HEAT AND THERMODYNAMICS

P384. A car's fuel tank is filled to 97% of its capacity with a volume V_G of gasoline. This process takes place at a temperature of $t = 0°$ C. The car is then transported by truck to a warm district, where the temperature is $t = 40°$ C. Is there a danger that the fuel will overflow the tank? (The volume expansion coefficients of the gasoline and the metal of the tank are $\gamma_G = 9 \times 10^{-4}\,°C^{-1}$ and $\gamma_T = 10^{-5}\,°C^{-1}$.

P385. The coefficient of thermal linear expansion of copper is $\alpha = 4 \times 10^{-6}\,°C^{-1}$, and its specific heat is $C = 0.386\ J\,g^{-1}\,°C^{-1}$. A square copper plate of side 10 cm and mass 100 g is heated from 0°C to 100°C.
   (a) – How much does the plate's area increase?
   (b) – How much heat does the plate absorb?

P386. A solid has thermal linear expansion coefficient α. Show that its volume expansion coefficient is $\gamma = 3\alpha$.

P387. A steel cube floats in a bath of mercury. What happens as the temperature rises? (Coefficient of linear expansion of steel $= \alpha_s = 1.2 \times 10^{-5}\,°C^{-1}$, coefficient of volume expansion of mercury $= \gamma_m = 1.8 \times 10^{-4}\,°C^{-1}$.)

P388. A heater is used to raise the temperature of water from $t_1 = 10°C$ to $t_2 = 38°$ C. It has to supply $V = 1\ m^3$ of hot water per hour. What is the minimum power that the heating element must supply? (The specific heat of water is $C_w = 4200\ J\,kg^{-1}\,°C^{-1}$.)

P389. An electric element of power $P = 1$ kW is used to heat a room of dimensions $4 \times 5 \times 2.5$ meters. Assuming that the efficiency of heating the air in the room is 75%, and that the air's heat capacity is $C_A = 1500\ J\,m^{-3}\,°C^{-1}$, how long does it take to heat the air in the room from $t_1 = 10°C$ to $t_2 = 20°C$?

P390. To prepare coffee, water has to be boiled starting from room temperature $t_1 = 15°C$. Assuming that the electric kettle is 50% efficient, how much does it cost to boil 1 liter of water if electricity costs 10 cents per kWh?

P391. A container holds a total mass $m = 1$ g of gas molecules, each with velocity $v = 600$ m s^{-1}. Find the total kinetic energy of the gas molecules.

P392. An ice cube of mass $m_I = 40$ g and temperature $t_I = -1°C$ is added to a glass of coke (mass $m_c = 200$ g) at room temperature $t = 20°C$. Neglecting any heat exchange between the drink (coke + ice) and its surroundings (glass + air), what will the temperature of the coke be once the ice has melted completely? The specific heat of ice is $C_I = 2310$ J kg^{-1} °C^{-1} and the latent heat of melting is $L_I = 3.36 \times 10^5$ J kg^{-1}. Assume that the coke has the same specific heat as water.

P393. Two animal species are similar in every respect except that every dimension of one is l times the corresponding dimension of the other. The species radiate excess heat from their surfaces and have plentiful supplies of the same type of food. By considering the heat balance of each species, explain why few small mammals are found in polar regions.

P394. A metal calorimeter has mass $m_c = 0.25$ kg and contains $m_w = 5$ kg of water, and the whole system is at a temperature $t_c = 10°C$. A block of mass $m_m = 10$ kg of the same metal as the calorimeter is removed from a container of boiling water and placed in the water inside the calorimeter. The insulated calorimeter–water–metal system reaches thermal equilibrium at a temperature of $t = 51°C$. Find the specific heat C_m of the metal.

P395. A bullet of mass $m = 10$ g is fired with velocity $v = 800$ m s^{-1} into a block of mass $M = 10$ kg of material with specific heat $C = 2000$ J kg^{-1} °C^{-1}. Assuming that all of the bullet's kinetic energy is used to heat the block (cf. P125, P126), by how much does its temperature rise?

P396. A copper calorimeter of mass $m_c = 125$ g contains $m_1 = 60$ g of water at a temperature of $t_1 = 24°C$. A mass $m_2 = 90$ g of hotter water with temperature $t_2 = 63°C$ is added, and the temperature of the calorimeter and water stabilizes at $t_3 = 45°C$. The calorimeter is perfectly insulated from its surroundings. Find the specific heat C_{cu} of copper in kcal kg^{-1}°C^{-1}.

P397. A mass $m_1 = 1$ kg of cold water at temperature $t_1 = 7°C$ is mixed with a mass $m_2 = 2$ kg of hot water at $t_2 = 37°C$. You may assume that no heat is exchanged with the surroundings, and that the total volume of water does not change. Find the temperature t of the mixture. Did the total internal energy of the water change? What was the total entropy change?

P398. A mass $m_g = 0.05$ kg of an ideal gas is held at a temperature of $t_1 = 0°C$ in a container of constant volume. The gas absorbs a quantity of heat $\Delta Q = 1.25 \times 10^5$ J, and as a result its pressure increases to three times its inital value. What is the final temperature t_2 of the gas? What is its specific heat at constant volume C_V (in J kg^{-1})?

P399. A glassful of water of mass $m_w = 0.25$ kg is boiled at atmospheric pressure and totally converted to steam. The latent heat of the water–steam transition is $L_w = 540$ kcal kg^{-1}. Find the change of entropy.

P400. A certain mass of gas is held at a pressure $P_1 = 2 \times 10^5$ N m^{-2} and occupies a volume $V_1 = 1$ m^3. The gas expands at constant pressure until its volume is doubled (i.e. $P_2 = P_1, V_2 = 2V_1$). It is then held at constant volume while its pressure is halved (i.e. $P_3 = P_2/2, V_3 = V_2$). A cyclic transformation (see Figure) is completed by an isobaric (constant pressure) compression ($P_4 = P_3$) to $V_4 = V_1$, followed by an isochoric (constant volume) transformation back to V_1, P_1. What is the work ΔW done by the gas? What is the absorbed heat ΔQ?

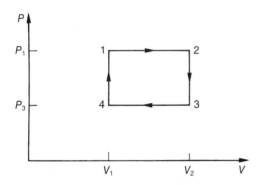

P401. A glass sphere of volume 7 liters contains air at 27°C and is attached to a pipe full of mercury as shown in the Figure. Initially the mercury is level with the bottom of the sphere in both arms of the tube, and the outside pressure is 760 mmHg. The air in the sphere is then heated so that the mercury level is raised by 5 mm in the outer arm. If the cross-sectional area of the pipe is 10 cm^2, what is the temperature of the air in the sphere?

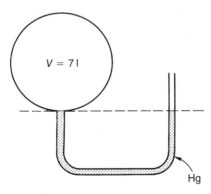

P402. An ideal gas of volume $V_1 = 400$ cm^3 and temperature $t_1 = 15°C$ expands adiabatically. As a result its temperature drops to $t_2 = 0°C$. If the gas has adiabatic index $\gamma = 1.4$, what is the volume V_2 of the gas after the expansion? The gas is then compressed isothermally until its pressure returns to the initial value (before expansion). What is its volume now?

P403. Five moles of an ideal monatomic gas expand adiabatically from an initial temperature $T_1 = 400$ K and pressure $P_1 = 10^6$ N m^{-2} to a final pressure $P_2 = 10^5$ N m^{-2}. Calculate the final temperature and the work done by the gas.

P404. The tires on a racing bicycle are generally inflated to pressures $P \approx 6\times$ atmospheric. When the valve is sharply depressed, ice forms around it. Why?

P405. Why does rain or snow tend to fall on the windward side of a mountain range? Why is there often a warm dry wind on the other side? (e.g. the Chinook on the eastern side of the Rockies.)

P406. Consider the balloon of P352 above. If instead of the temperature being fixed, the monatomic gas inside the balloon expands adiabatically, show that its maximum radius is smaller than in P352. Why?

P407. A certain mass of ideal gas, with constant-volume specific heat $C_V = 0.6$ J mol^{-1} K^{-1}, is cooled at constant pressure $P_0 = 10^5$ N m^{-2}. As a result its volume decreases from $V_1 = 1$ m^3 to half of this value. Find the amount of heat lost by the gas in this process.

P408. Two moles of an ideal monatomic gas expand isobarically (i.e. at constant pressure) from an initial volume $V_1 = 0.03$ m^3 to a final volume $V_2 = 0.07$ m^3. The pressure throughout is $P = 1.52 \times 10^5$ N m^{-2}. Calculate the initial and final temperatures T_1, T_2 of the gas, the total amount of heat Q absorbed in the process, and the change ΔS in the entropy of the gas.

P409. Two solid bodies of equal masses m and temperatures T_1 and $T_2 = 2T_1$ are brought into contact. If their heat capacities are C_1 and $C_2 = 1.5C_1$, what is their common temperature, T, when they reach thermal equilibrium? Find the entropy change ΔS for each body, and show that the total entropy of the system has increased. Express your results in terms of T_1, C_1 and m.

P410. A mass $m = 0.16$ kg of molecular oxygen (O_2) at a temperature $T_1 = 300$ K and a pressure $P_1 = 1$ atm $= 10^5$ N m^{-2} is adiabatically compressed to a pressure $P_2 = 10$ atm. Calculate the final volume V_2 and temperature T_2 of the oxygen. What quantity of work ΔW is performed in the compression, and what is the change ΔU of internal energy?

P411. The volume of an ideal gas is doubled in a quasistatic isothermal process. Find the change in the pressure P, temperature T, internal energy U, and entropy S. Express the changes $\Delta P, \Delta T, \Delta U, \Delta S$ in terms of the initial values P_0, T_0, U_0, S_0, V_0 and n, the number of molar masses of gas. (Use the formula $\Delta W = nRT \ln(V/V_0)$ for the work done by an isothermal ideal gas in expanding from volume V_0 to V).

P412. A heat pump is used to heat a house by absorbing a certain heat quantity Q_2 from the outside air (temperature T_2) and supplying a quantity of heat Q_1 to the house (temperature T_1), with $T_2 < T_1$. The machine works cyclically, and on each cycle a quantity W of work is performed (by an electric motor). Find the relation between Q_1, Q_2, and W. If the machine is completely efficient, how much heat will be supplied to a house at $T_1 = 17\,^\circ\text{C}$ with an outside temperature $T_2 = -5\,^\circ\text{C}$ for every joule of output from the electric motor?

P413. N gas molecules, each with mass m, are confined in a cube of volume V. Show that the pressure on the walls is

$$P = \frac{Nmv^2}{3V},$$

where v is the root-mean-square (rms) speed of the molecules, defined as

$$v^2 = \frac{1}{N}(\Sigma v_x^2 + \Sigma v_y^2 + \Sigma v_z^2).$$

P414. Three gas molecules have speeds $v_i = 1, 3$ and $10\ \text{m s}^{-1}$ in the same direction. Find (a) their average speed and (b) their rms speed v, where

$$v^2 = \frac{1}{3}\Sigma v_i^2.$$

P415. Show that the rms speed of molecules of a gas is

$$v = \left(\frac{3kT}{\mu m_H}\right)^{1/2}$$

where T is the absolute temperature, R the gas constant, and μ the mean molecular mass.

P416. Find the rms speed of oxygen molecules (mean molecular mass $\mu = 32$) and hydrogen molecules ($\mu = 2$) at room temperature ($T = 300$ K).

P417. A bottle of perfume is opened in one corner of a large room. Show that typical molecular rms speeds do not give a good estimate of how soon you would expect to notice the scent in a distant part of the room? Why not?

P418. Show that the specific heat per unit mass at constant volume for a monatomic gas is $3k/2\mu m_H$.

One kilojoule of energy is required to raise the temperature of a certain mass of helium gas ($\mu = 4$) through 30 K. How much is needed to raise the temperature of the same mass of argon ($\mu = 40$) by the same amount?

Explain this result in terms of microscopic properties of the two gases. (Both helium and argon are monoatomic.)

P419. The gas in a cylinder is adiabatically compressed by a piston. By considering microscopic processes, explain qualitatively why its temperature and pressure rise.

P420. A box containing gas is weighed on a scale. Most of the gas molecules are not in contact with the base. Why does the scale nevertheless register the weight of the gas as well as the box?

P421. The escape speed from the Earth (see S187) is $v_{esc} = 11.2$ km s^{-1}. At what temperature would the following gases tend to escape from the Earth's atmosphere: nitrogen ($\mu = 28$), oxygen ($\mu = 32$) and hydrogen ($\mu = 2$)?

■ LIGHT AND WAVES

P422. The base angles of a triangular glass prism are $\alpha = 30°$, and its refractive index is $n = 1.414$ (see Figure). Parallel light rays A and B are normally incident on its base. What is the angle between the two emergent rays?

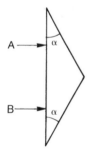

P423. A light ray is incident on side AB of an equilateral triangular prism at angle α (see Figure). If $\alpha < 90°$ some of the light emerges through side AC, but if $\alpha \geq 90°$, no light emerges through this side. Calculate the refractive index n of the prism glass.

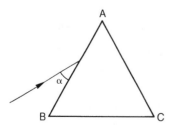

P424. A light ray is incident at $40°$ on a glass plate of refractive index $n = 1.3$ and width $h = 1$ cm, and emerges from the other side of it. Find the linear displacement of the light ray caused by refraction.

P425. A swimming pool is illuminated by an underwater point source of light. Viewed from above the water at a horizontal distance $d = 1$ m the light is seen at an angle $\theta_2 = 30°$ (see Figure). How deep is it? (Refractive index n of water $= 1.3$.)

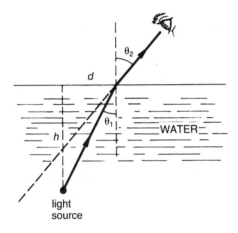

P426. A light ray is incident on the end of a straight optical fiber at angle θ_1 and enters the fiber at angle θ_2 (see Figure). If the refractive index of the fiber is n, what is the maximum value of θ_1 such that the ray remains within the fiber? (Express your answer in terms of n.)

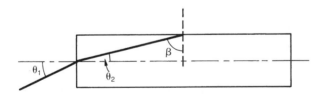

P427. A beam of white light is incident at angle $\alpha = 30°$ on a water droplet with refractive index $n = n(\lambda)$ given as a function of wavelength λ (see Figure). As

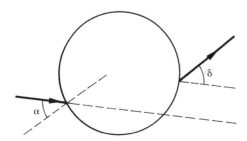

the ray emerges from the far side of the droplet it has been deflected through an angle δ from its original path. Calculate δ as a function of λ. If $n(\lambda)$ is such that $n = 1.53$ for blue light and $n = 1.52$ for red light, by how much will the corresponding deflections differ?

P428. A candle is placed a distance $s = 1.5$ m along the axis of a convex spherical mirror of curvature radius $R = 1$ m (see Figure). Find the position, nature, and magnification of the image. Draw a schematic ray diagram.

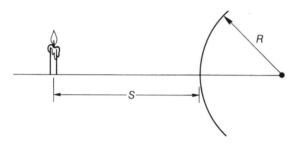

P429. An object is on the axis of a concave spherical mirror of curvature radius $R = -2$ m. Its image is twice the object size and appears in front of the mirror. Find the positions of the object and image, and supply a ray diagram.

P430. An object is placed at a distance $s = R/4$ from a concave spherical mirror of curvature radius R. Find the position and nature of the image. Draw a ray diagram.

P431. An experimenter wishes to produce an image of the coil of an electric lamp on a wall, with the aid of a spherical mirror. The coil is a distance $s = 0.1$ m from the mirror, which is itself $d = 3$ m from the wall. What kind of mirror (concave or convex, and what radius of curvature) should the experimenter use? What is the image size if the coil is $h = 0.5$ cm long? Give a ray diagram.

P432. Calculate the focal lengths of the following thin glass ($n = 1.5$) lenses:
    (a) – biconvex, with radii $R_1 = 1$ m, $R_2 = 1.3$ m,
    (b) – biconcave, with the same radii,
    (c) – concave–convex, with the same radii,
    (d) – convex–concave, with the same radii,
    (e) – one flat surface, the other convex with $R_2 = 1.3$ m.

P433. A converging lens with focal length $f = 10$ cm is used to observe an insect of size h. Find the position, nature and size (in terms of h) of the image if
    (a) – the insect is $s = 5$ cm from the lens, and
    (b) – the insect is $s = 15$ cm from the lens.
Give a ray diagram in each case.

P434. A bright object is placed a distance $s = 1$ m from a converging lens of focal length $f = 0.5$ m. A plane mirror is placed perpendicular to the optical axis on the opposite side of the lens. How many images are formed? Determine whether each image is real or virtual, and upright or inverted. Check your conclusions by means of drawings.

P435. A point light source is a height $h = 50$ cm above a table. An experimenter wishes to obtain a sharp image of the source at the table, using a converging lens of focal length $f = 8$ cm. At what height x should she place the lens?

P436. Show that the thin lens formula can be rewritten as

$$pp' = f^2,$$

where p, p' are the distances of the object and image from the first and second focal points.

P437. Two thin lenses of focal lengths f_1, f_2 are placed in contact. Show that they are equivalent to a thin lens with focal length f given by

$$\frac{1}{f} = \frac{1}{f_1} + \frac{1}{f_2}.$$

P438. Two lenses of power $P_1 = 2$ diopters and $P_2 = 0.5$ diopters are placed in contact. What is the power of the combined lens?

P439. An optical doublet is formed from two lenses A, B made of glass of different refractive indices n_A, n_B. Lens A has two convex sides of radius of curvature R, and lens B has one flat side and one concave side of radius of curvature R. Derive an expression for the power of the doublet.

Both refractive indices vary slightly with wavelength as follows: $n_A = 1.50, 1.51, 1.52$ at red, yellow, and blue respectively, while $n_B = 1.60, 1.62, 1.64$ at the same wavelengths. Show that the doublet has constant power at all three wavelengths.

P440. A simple camera has a converging lens of focal length $f = 5$ cm and is used to record sharp images of distant objects on film. If instead the objects are $s = 1$ m from the lens, by how much must the distance between the lens and the film be changed?

P441. Show that, except for extreme closeups, the magnification of a camera lens is approximately proportional to the focal length of its lens. How are different magnifications achieved in practice? Does this affect the field of view?

P442. A photographer uses a camera with an $f/8$ lens and obtains a good picture with an exposure of 0.02 s. The diaphragm is now stopped down to $f/16$ and the lighting conditions remain the same. What exposure is now required?

P443. By changing the radii of its converging lens, and thus its focal length, the human eye is able to produce a sharp image on the retina (at a fixed distance from the lens) of objects at any distance from a certain minimum (the "least distance of distinct vision", or "near point") up to infinity. If the near point is a distance $d_n = 25$ cm from the eye's lens, and the retina is 2.5 cm behind the lens, by what factor must the eye muscles be able to change the lens's focal length?

P444. A normal human eye can produce a sharp image of an object at any distance beyond a near point (about 25 cm, see the previous problem) all the way out to infinity. A certain person has an eye with a normal near point, but is unable to see clearly objects beyond a far point at $d_f = 1$ m. How can her vision be corrected?

P445. A man has a near point at $d_n = 0.6$ m from his eyes. What power glasses will bring his near point to $d_n' = 0.25$ m?

P446. The human eye can distinguish point objects down to angular separations $\theta_0 \approx 5 \times 10^{-4}$ rad ($\approx 0.03° \approx 1.7'$). If a person has a near point $d_n = 25$ cm, what is the size of the smallest detail that he can pick out?

P447. A person with a near point $d_n = 25$ cm uses a converging lens with a power of 10 diopters to view a very small object. Where must the object be placed with respect to the lens for best results, and how large is the angular magnification?

P448. A microscope has an objective lens of focal length $f_1 = 1$ cm and an ocular lens of focal length $f_2 = 5$ cm. What is its angular magnification? It is used to view a specimen at distance $s_1 = 1.1$ cm from the objective. What is the size of the smallest detail that can be observed by a normal eye using the microscope?

P449. The focal length of a certain astronomical reflecting telescope is $f = 15$ m. The image is viewed through an eyepiece of focal length $f_e = 3$ cm. What is the angular magnification? Why would it be difficult to build a refracting telescope of the same magnification?

P450. A wave is described by the formula

$$y(x, t) = 0.1 \sin\left[2\pi\left(\frac{t}{0.01} - \frac{x}{5}\right)\right],$$

where y and x are in meters and t is in seconds. What are the amplitude A, wavelength λ, phase velocity v_ϕ and frequency ν?

P451. A sinusoidal wave of frequency $\nu = 10^3$ Hz has phase velocity $v_\phi = 500$ m s^{-1}. What is its wavelength λ? Find the distance between any two points with a phase difference $\Delta\phi = \pi/6$ rad at any given time. At a fixed point, by how much does the phase change over a time interval $\Delta t = 10^{-4}$ s?

P452. A car driven by a physicist is stopped by a policeman who claims that it passed a traffic light on red. The physicist tries to convince the policeman that the light appeared as yellow because of the Doppler effect. Is the policeman justified in giving the physicist a speeding ticket? (The wavelengths of red and yellow light are 6900 Å, 6000 Å.)

P453. A uniformly moving train sounds its horn as is passes a stationary observer. The observer hears the horn note a factor 1.2 lower in frequency after it passes than before. What is the train's speed (speed of sound v_s in air $= 330\,\mathrm{m\,s^{-1}}$)?

P454. A car horn moving at $v = 40\ \mathrm{m\ s^{-1}}$ towards a static pedestrian emits a sound wave of frequency $\nu_0 = 500$ Hz. The sound speed is $v_s = 340\ \mathrm{m\ s^{-1}}$.
(a) – What is the wavelength λ emitted by the horn?
(b) – At what frequency ν does the pedestrian hear the horn?

P455. An astronomer uses a telescope and spectrograph to observe a set of absorption lines in the spectrum of a star. All of them are shifted slightly to the red compared with the same lines in the Sun. In particular the Hα line ($\lambda_0 = 6562$ Å in the Sun) appears at $\lambda = 6563$ Å. What can you conclude about the motion of the star?

P456. An astronomer uses a telescope and spectrograph to observe the spectrum of one star of a binary system (two stars orbiting about their common center of mass). If he continues to observe for long enough, what will he notice?

P457. Two identical sound sources A and B are 1 m apart under water and emit sound waves of frequency $\nu = 3500$ Hz in phase with each other. A microphone is placed on a line parallel to AB at a distance $L = 1000$ m from AB. Where should it be positioned so that the sound intensity is a local maximum? (Speed of sound in water $= 1500\ \mathrm{m\ s^{-1}}$.)

P458. In the arrangement of the previous problem, the microphone is placed at position $x = 474.4$ m. The emitted frequency is now adjusted in the range $2500 \le \nu \le 5500$ Hz. What value should it take so that the microphone now detects *zero* sound intensity?

P459. A Young's double slit experiment is performed using light of wavelength $\lambda = 5000$ Å, which emerges in phase from two slits a distance $d = 3 \times 10^{-5}$ cm apart. A transparent sheet of thickness $t = 1.5 \times 10^{-5}$ cm is placed over one of the slits. The refractive index of the material of this sheet is $n = 1.17$. Where does the central maximum of the interference pattern now appear?

P460. In a two-slit interference pattern (Young's experiment) the slits are a distance $d = 0.3$ mm apart. A screen is placed at $L = 1$ m from the slits, which are illuminated by light of one wavelength only (monochromatic beam). In the interference pattern on the screen the 8th maximum is a distance $D = 1.46$ cm from the principal maximum. Find the wavelength λ of the light in nanometers.

P461. A spectrometer makes use of a grating with 5000 lines cm^{-1}. At what angles will maxima of light of wavelength $\lambda = 6563$ Å appear? If white light (4000 Å$\leq \lambda \leq 7000$ Å) is analyzed by the spectrometer, over what range of angles do the second- and third-order interference patterns overlap?

P462. A laser beam of light at $\lambda = 6870$ Å passes through a slit of width $D = 10^{-4}$ cm. In what directions is the intensity zero? What happens if D is doubled?

P463. A parallel beam of light of wavelength $\lambda = 7000$ Å passes through a narrow slit in an opaque screen. It produces a central intensity maximum of width $\Delta z = 1.4$ cm (between the zeros on each side of the maximum) on a second parallel screen $L = 1$ m from the first. What is the width of the slit?

P464. A thin uniform layer of oil of refractive index $n = 1.25$ lies on a perfectly reflecting flat surface. A monochromatic light beam of wavelength λ (in air) is normally incident on the oil. In terms of λ, for what thickness d of oil will the reflected intensity be (a) a minimum, (b) a maximum?

P465. A mob official wishes not to be seen through the windows of her Mercedes in daylight (dominant wavelength λ). The refractive index of the car's window glass is $n_g = 1.4$. To minimize light transmission, the mob's engineer has the windows coated with a thin layer of optical paint with refractive index $n_s = 1.5$. The width of the layer is chosen to be $d = 7\lambda_p/2$, where λ_p is the light wavelength in the paint. Speculate on the engineer's fate.

P466. A soap film (refractive index $n = 1.3$) is illluminated by monochromatic light of wavelength $\lambda = 5200$ Å. Initially the film has thickness d_0 and its transparency is maximal, but it is gradually stretched until its thickness reaches d_1 and its transparency reaches a minimum. Find the possible values of d_0 and d_1.

■ ATOMIC AND NUCLEAR PHYSICS

P467. Calculate the de Broglie wavelength of electrons whose speed is $v_e = 10^7$ m s^{-1}. What experiment could one perform to distinguish between a beam of such electrons and a beam of photons having the same wavelength?

P468. In a certain metal, the binding energy of electrons (the work function) is $B = 3 \times 10^{-19}$ J. The metal is illluminated by a monochromatic beam of light of wavelength λ. What is the maximum value of λ such that photo-electrons are emitted? If $\lambda = 4.4 \times 10^{-7}$ m, calculate the maximum kinetic energy E_{max} of the photoelectrons and the stopping potential V_s. How do these two results depend on the *intensity* of the beam?

P469. When illuminated by monochromatic light of wavelength $\lambda = 5500\,\text{Å}$, a certain metal emits electrons with a maximum energy of $E_e = 1.02$ eV. When the metal is illuminated by monochromatic light of wavelength $\lambda' = 4800\,\text{Å}$, the maximum electron energy is $E'_e = 1.35$ eV. Find the value of Planck's constant h from these data. Can such an experiment be performed using any metal? Explain your answer.

P470. Calculate the number of photons emitted per second by a radio transmitter broadcasting at a frequency of $\nu = 1\,\text{MHz}$ with power $P = 10\,\text{kW}$.

P471. In a certain experiment, the position of an electron is determined to an accuracy $\Delta x = 10^{-9}$ m. Assuming that the electron is non-relativistic, what is the most accurate knowledge we can hope to have about its velocity in this experiment?

P472. Find the energy (in both joules and electron volts) and momentum of an X-ray photon of frequency $\nu = 5 \times 10^{18}$ Hz.

P473. The electron current in an X-ray tube is $I = 16$ mA, and the potential difference is $\Delta V = 12,000$ V. What is the shortest wavelength of the emitted photons? How many electrons hit the anode per second?

P474. What is the de Broglie wavelength of the Earth moving in its orbit? Using the Bohr model for the Sun–Earth system, find the quantum number n of the orbit. (You may assume that the Earth has mass $M_e = 6 \times 10^{24}$ kg and moves in a circular orbit of radius $R = 1.5 \times 10^{11}$ m.) What can you say about the applicability of quantum versus classical mechanics in this case?

P475. Electrons are accelerated in a cathode ray tube by a potential difference of $V_0 = 5000$ V.
   (a) – What is the de Broglie wavelength of the electrons?
   (b) – What is the shortest wavelength of photons emitted by the anode when electrons hit it?

P476. A photon of wavelength $\lambda = 0.2\,\text{Å}$ encounters a stationary electron and is scattered directly backwards. Calculate the final wavelength λ' of the photon, and the electron's kinetic energy E'_e after the collision.

P477. A gamma ray of wavelength $\lambda_1 = 0.0048$ nm is Compton scattered at an angle θ from an electron at rest. After the scattering, the magnitudes of the photon and electron momenta are equal. Find the angle θ and the wavelength λ_2 of the photon after scattering.

P478. The quantization condition of Bohr's theory of the hydrogen atom is $m_e v_n r_n = n\hbar$, where v_n, r_n are the velocity and radius of the nth electron orbit. Show that this is equivalent to requiring the circumference of the orbit to be n times the electron's de Broglie wavelength.

P479. Use Bohr's quantization condition (see previous question) and classical mechanics to find the total energy of the nth orbit in the hydrogen atom. Express the ground state energy in terms of physical constants.

P480. An electron collides with a gas of atomic hydrogen, all of which is in the ground state. What is the minimum energy (in eV) the electron must have to cause the hydrogen to emit a Balmer line photon?

P481. A hydrogen atom in the $n = 4$ state makes a transition to the ground state, emitting one photon. Calculate the wavelength of the emitted photon and the recoil velocity of the atom.

P482. Calculate the energy of levels $n = 100$ and $n = 1000$ in the Bohr model of the hydrogen atom. What can you say about the binding energy of the electron in these orbits? Describe the spectrum of radiation emitted when such states make a transition to a given low-lying level.

P483. Use the Bohr model of the hydrogen atom to show that when an electron jumps from the level n to level $n - 1$ the frequency of the emitted photon is close to the electron rotation frequency (in Hz) if n is very large.

P484. Figure 1 represents the energy levels of a certain atom. If a gas of such atoms is irradiated by a beam of white light, what absorption lines are expected in the spectrum, when the experiment is viewed along the beam axis (see Figure 2)?

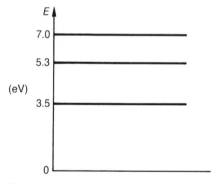

Fig 1

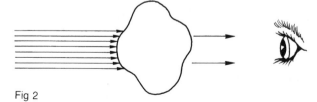

Fig 2

P485. An atom of singly ionized helium has a single electron, whose energy levels are given by an expression similar to that of a hydrogen atom, i.e.

$$E_n = -\frac{4E_0}{n^2}, \tag{38}$$

where $E_0 = 13.6$ eV. What is the minimum energy required to ionize a helium atom completely?

A beam of electromagnetic radiation has a continuous spectrum extending between $\lambda_{\text{low}} = 240$ Å and $\lambda_{\text{high}} = 500$ Å; it is incident on an ensemble of singly ionized helium atoms, which are all in the ground state. Calculate the wavelengths of the absorption lines involving transitions from the ground state seen if the experiment is viewed along the beam axis. How many different emission lines will be seen in this case? How many are seen if the experiment is viewed from the side?

P486. A sample of sodium containing a certain concentration of the $_{11}\text{Na}^{24}$ isotope is prepared. After 60 hours this concentration has fallen to 7% of its original value. Calculate the half-life $t_{1/2}$ of $_{11}\text{Na}^{24}$.

P487. An isotope of iron ($Z = 26, A = 59$) undergoes beta decay into a stable isotope of cobalt. Find Z and A for the cobalt isotope. In 30 days the number of radioactive iron atoms in a certain sample decreases from $N_1 = 10^{20}$ to $N_2 = 6.25 \times 10^{19}$. What is the half-life of the iron isotope?

P488. The half-lives of the two uranium isotopes U^{238}, U^{235} are known to be $t_{1/2}(\text{U}^{238}) = 4.5 \times 10^9$ yr, $t_{1/2}(\text{U}^{235}) = 7.1 \times 10^8$ yr. If the Earth was formed with equal amounts of the two isotopes, estimate its current age, given that uranium ores are now 99.29% U^{238} and 0.71% U^{235} by number.

P489. The radioactive element ^{14}C decays by beta emission. In a living organism the activity of ^{14}C (i.e. the number of decays per minute per gram) is known to be 15.3. In a certain archaeological excavation a human bone is found in which the activity is 1.96. The half-life of ^{14}C is $t_{1/2} = 5568$ y. Estimate the age of the bone.

P490. When a helium nucleus is formed from two deuterium nuclei an energy of 23.8 MeV is released. In the fission of U^{235} an energy of approximately 200 MeV is released. Compare the total amount of energy released in the fusion of 1 g of deuterium with that released in the fission of 1 g of U^{238}.

■ RELATIVITY

P491. A body moves uniformly relative to an observer, who measures its length and finds a value $l = l_0/2$, where l_0 is its proper length. What is the velocity v of the body? A clock moving with the body measures a time interval $\tau_0 = 1$ s between two events. What does the observer measure for this interval?

P492. An electron moves so that its total energy is twice its rest-mass energy. What is its velocity? At what velocity is its momentum mc, where m is its rest-mass?

P493. A certain elementary particle lives only a time $\tau_0 = 5$ s before disintegrating. What velocity must the particle have if it is to reach the Earth from the Sun (distance $l = 1.5 \times 10^{11}$ m) before disintegrating?

P494. A spaceship S_1 moves with uniform velocity $v = 0.99c$ with respect to a space station S_2. The clocks in S_1 and S_2 are synchronized at zero hours as the spaceship passes the space station. The captain of S_1 sends a radio signal to S_2 when his clock reads 1.00 hr. What will S_2's clock read when the signal reaches it?

P495. A spaceship moves with velocity $v_s = 0.6c$ directly towards a space station. It fires a missile at the station with velocity $v_m = 0.5c$ with respect to itself. What is the missile's velocity with respect to the station? Repeat the calculation for the case $v_s = 0.001c$. Compare your results in both cases with the answer given by the non-relativistic velocity addition formula: does the latter provide a good approximation in either case?

P496. A particle of mass m moves with velocity $v = 0.8c$ in the laboratory frame and collides with an identical stationary particle, combining with it to create a new single particle of mass M and velocity V. Find M, V.

P497. An electron and a positron (each of mass $m_e = 9.1 \times 10^{-31}$ kg) collide with velocities $\pm v = \pm 0.6c$ in the laboratory frame, and gamma radiation is emitted. Show that more than one photon must be emitted. If exactly two photons are emitted show that they must move in opposite directions and have equal energies E. Calculate E and the corresponding photon wavelength λ.

P498. A cosmic-ray source moves with velocity $v_s = 0.6c$ away from the Earth. In its rest frame it emits protons with energy $E = 2000$ MeV in all directions. Calculate the speed v_p in the source frame and v_p' in the Earth's frame of a proton emitted towards the Earth. How long (in the Earth's frame) will it take for a proton to reach the Earth if emitted at a distance $l = 10^{15}$ km? What is the corresponding time in the proton's frame? ($m_p = 1.67 \times 10^{-27}$ kg.)

P499. An alien spaceship moves with constant velocity $v = 0.6c$ relative to the Earth. It passes the Sun at a certain point on its way to the Earth (you may neglect the Earth's motion about the Sun in this problem). How long does the Sun–Earth journey take according to a terrestrial observer? How long do the aliens measure the trip as taking? (Earth–Sun distance $l = 1.5 \times 10^8$ km.)

P500. When a spaceship passes the Earth, an alien aged 20 Earth-years falls in love with a terrestrial student whom she sees on her monitor screen. At the time the student is also exactly 20 years old. The relationship is discouraged by the alien authorities and the spaceship continues to move at constant speed $v = 0.998c$. After one year (spaceship time) the alien is able to send a radio message to the student. How old is the student when the message arrives at Earth?

PART **TWO**

SOLUTIONS

CHAPTER **ONE**

MECHANICS

☐ **STATICS**

S1. Choosing the origin at the center of the Sun and the x-axis along the Sun–planet direction, we have for the Earth–Sun system

$$x_{CM} = \frac{0 + M_e d_e}{M_\odot + M_e} \approx \frac{M_e}{M_\odot} d_e = 3 \times 10^{-6} \times 1.5 \times 10^{11} = 4.5 \times 10^5 \text{ m},$$

which is well inside the Sun, i.e. $x_{CM} \ll R_\odot$.

For the Jupiter–Sun system

$$x_{CM} = \frac{0 + M_J d_J}{M_\odot + M_J} \approx \frac{M_J}{M_\odot} d_J = 10^{-3} \times 1.4 \times 10^{12} = 1.4 \times 10^9 \text{ m}.$$

This is outside the Sun (about $2R_\odot$ from its center).

S2. We choose the origin of coordinates at the center of the hoop and the x-axis along the shaft (see Figure). The positions $(x_1, y_1), (x_2, y_2)$ of the centers

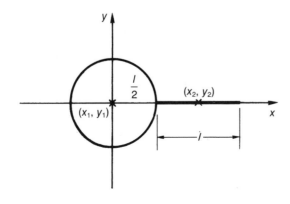

of mass of the hoop and shaft are obviously given by $x_1 = 0$, $y_1 = 0, x_2 = l, y_2 = 0$, so the center of mass of the entire racket is given by

$$x_{CM} = \frac{m_1 x_1 + m_2 x_2}{m_1 + m_2} = ml \cdot \frac{1}{2m} = \frac{l}{2}$$

with $y_{CM} = 0$. The center of mass is where the shaft joins the hoop. This is obvious by symmetry, as the hoop and shaft have equal masses and their centers of mass are equally spaced about that point.

S3. In calculating x_{CM} we have to add a mass $m_3 = m/2$ with coordinates $x_3 = -l/2, y_3 = 0$ to the expression in S2 above. This gives

$$x_{CM} = \frac{-ml/4 + ml}{5m/2} = \frac{3}{10}l.$$

The new center of mass is inside the hoop, a distance $l/5$ from the point where the shaft joins it.

S4. The center of mass of a triangle of uniform density and thickness is at its centroid, i.e. the intersection of the medians (see Figure). The centroid divides each of the medians in the ratio of 2:1, so the center of mass of the eaten slice is at a position $2r/3$ from the center of the pizza. Choosing the origin of coordinates at the center of the pizza and the x-axis along the symmetry line of the slice, the center of mass of the full pizza lies at $x = 0$, while those of the slice and pizza minus slice lie at $x_s = 2r/3$ and x_e, respectively. Thus

$$0 = \frac{m_s x_s + m_e x_e}{m_s + m_e},$$

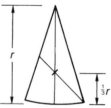

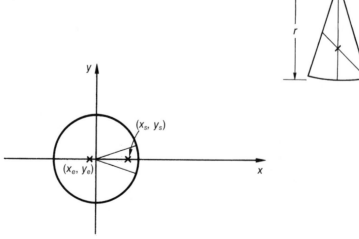

where m_s, m_e are the masses of the slice and pizza minus slice. Since the pizza is uniform $m_s = (20/360)m = 0.056m$, and $m_e = m - m_e = 0.944m$, so

$$x_e = -\frac{m_s}{m_e}x_s - \frac{0.667r \times 0.056}{0.944} = -0.04r.$$

Thus the balance point is shifted away from the original center of the pizza by only 4% of the radius.

S5. The ballast lowers the center of mass. This makes the boat more stable: if the center of mass is too high, the boat may even capsize.

S6. The board is placed across the two scales as shown in the Figure, and the person lies on it. The extra weights W_1, W_2 registered by the scales are noted. If the scales are a distance d apart and the center of mass (CM) is a distance a from the top of the left-hand scale, requiring $\Sigma M_O = 0$ about the CM gives $W_1 a = W_2(d - a)$, i.e.

$$a = \frac{W_2 d}{W_1 + W_2}.$$

We may regard this as the z coordinate of the CM.

The process is then repeated with the person standing facing a particular direction, and then facing at right angles to it, giving also the x, y coordinates of the CM.

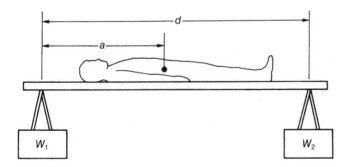

S7. The forces acting on the body are its weight W, the static frictional force f_s and the normal reaction force N of the plane (see Figure). The latter two are exerted by the inclined plane. The weight is a result of the Earth's gravity. To calculate the force we choose a Cartesian coordinate system with the y-axis normal to the plane and the x-axis down it. In equilibrium, as here, we have $\Sigma F_x = \Sigma F_y = 0$, or

$$W \sin \theta - f_s = 0, \tag{1}$$

$$N - W \cos \theta = 0. \tag{2}$$

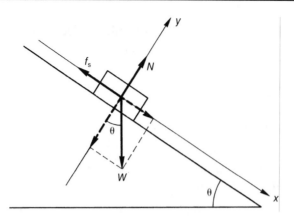

With $W = mg$ this gives $f_s = mg \sin \theta = 5 \times 9.8 \times \sin 30° = 24.5$ N and $N = 5 \times 9.8 \times \cos 30° = 42.4$ N. The maximum value the frictional forces can have is $f_s^{\max} = N\mu_s = 42.4 \times 0.6 = 25.4$ N, and this exceeds the actual value of f_s we have calculated above, which prevents the body sliding. In general, equations (1) and (2) show that $f_s = mg \sin \theta$ and $N\mu_s = \mu_s mg \cos \theta$, so that equilibrium is possible for $f_s \leq N\mu_s$, i.e. $mg \sin \theta \leq \mu_s mg \cos \theta$, or $\mu_s \geq \tan \theta$. Here $\tan 30° = 0.58 < 0.6$, as required.

S8. Choosing the origin of coordinates at the mass m with the x, y axes respectively horizontal and vertical, the conditions for equilibrium are $\Sigma F_x = 0, \Sigma F_y = 0$. With T_1, T_2 the tensions in the strings we have (see Figure)

$$T_1 \cos \alpha + T_2 \cos \beta - mg = 0$$

$$T_1 \sin \alpha - T_2 \sin \beta = 0.$$

The second equation can be rewritten as $T_1 = T_2 \sin \beta / \sin \alpha$, allowing us to eliminate T_1 from the first equation:

$$T_2[\cos \alpha \sin \beta + \sin \alpha \cos \beta] = mg \sin \alpha,$$

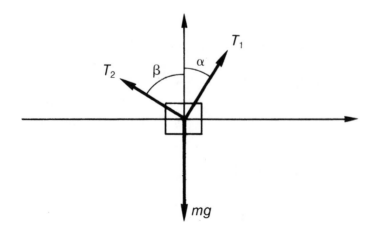

so using the trigonometric identity for $\sin(\alpha + \beta)$,

$$T_2 = \frac{mg \sin \alpha}{\sin(\alpha + \beta)}.$$

The relation between T_1 and T_2 shows that

$$T_1 = \frac{mg \sin \beta}{\sin(\alpha + \beta)}.$$

Now using $\alpha = 45°, \beta = 60°$ gives $T_1 = 0.897mg, T_2 = 0.732mg$. The equilibrium of the two vertically hanging weights requires $T_1 = m_1 g, T_2 = m_2 g$, and thus $m_1 = 0.897m = 8.97$ kg, $m_2 = 0.732m = 7.32$ kg.

S9. Let the string make an angle α to the wall. As the wall is smooth, there is only a normal reaction force N between it and the ball. Taking the x and y axes horizontal and vertical, the equilibrium conditions $\Sigma F_x = 0, \Sigma F_y = 0$ become (see Figure)

$$N - T \sin \alpha = 0 \tag{1}$$

$$T \cos \alpha - mg = 0. \tag{2}$$

Then from (2), $T = mg/\cos \alpha = mg \sec \alpha$. Since $\tan \alpha = r/h = 1/\sqrt{3}$, the identity $\sec^2 \alpha = 1 + \tan^2 \alpha$ shows that $\sec \alpha = 2/\sqrt{3}$ so that $T = 2mg/\sqrt{3}$. Now (1) shows that $N = mg \tan \alpha = mg/\sqrt{3}$.

If the wall is rough, (2) above becomes instead

$$\mu_s N + T \cos \alpha = mg. \tag{3}$$

Eliminating N between (3) and (1) gives

$$T = \frac{mg}{\mu_s \sin \alpha + \cos \alpha},$$

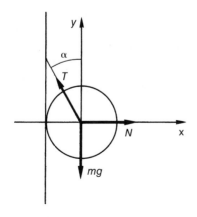

and (1) shows that

$$N = \frac{mg \sin \alpha}{\mu_s \sin \alpha + \cos \alpha}.$$

As can be seen, both T and N are reduced by nonzero μ_s: the effect of friction is to help support the sphere, reducing the required tension in the string and thus the normal force on the wall.

S10. Taking the x and y axes horizontal and vertical, we see from the Figure that the horizontal equilibrium condition $\Sigma F_x = 0$ is satisfied by symmetry. With α the angle of the two rope sections to the horizontal, the vertical equilibrium condition $\Sigma F_y = 0$ is

$$2T \sin \alpha - mg = 0. \tag{1}$$

The length of the stretched rope is $l = l_0/\cos \alpha$ (each section is stretched by a factor $1/\cos \alpha$), so that

$$T = \kappa(l - l_0) = \kappa l_0 \left(\frac{1}{\cos \alpha} - 1 \right).$$

Thus substituting for T in (1) shows that

$$2\kappa l_0 (1 - \cos \alpha) \tan \alpha = mg. \tag{2}$$

The critical (maximum) angle α_c has $\tan \alpha_c = h/l_0 = 1/6$, so that $\alpha_c = 9.46°$, and $\cos \alpha_c = 0.986$. From (2) we thus find that κ must have at least the value $\kappa_c = mg[2l_0 \tan \alpha_c (1 - \cos \alpha_c)]^{-1} = 60 \times 9.8(2 \times 6 \times 1/6 \times 0.014)^{-1} = 2.1 \times 10^4$ N m^{-1}. If the performer hangs vertically from the rope, we must have the equilibrium condition

$$mg = T = \kappa(l - l_0),$$

so that the extension of the rope is $l - l_0 = mg/\kappa = 60 \times 9.8/2.1 \times 10^4 = 0.028$ m, i.e. less than 3 cm. The big difference from the earlier case results from the fact that there the rope was almost horizontal, so that a much larger tension was needed to balance the performer's weight.

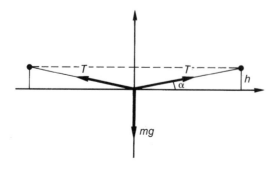

S11. This problem is a particular case of P8, with now $\alpha = \beta, T_1 = T_2 = T$. Using the relation for T_1 or T_2 in S8

$$T = mg \frac{\sin \alpha}{\sin 2\alpha} = \frac{mg}{2 \cos \alpha}.$$

Thus

$$\cos \alpha = \frac{mg}{2T}.$$

A horizontal wire would have $\alpha = 90°$ or $\cos \alpha = 0$. For $m \neq 0$ this is impossible, however large T becomes. For $T = 100mg$ we find $\alpha = 89.7°$, i.e. the wire makes an angle $0.3°$ to the horizontal.

The wire must always sag slightly in order to balance the weight of the mass. Since the wire itself always has mass, it can never be stretched completely horizontal. This effect can be seen easily by looking at a tennis net.

S12. The vertical equilibrium of the hanging weight, $\Sigma F_y = 0$, gives $T = W$, where T is the tension in the cord. Using $\Sigma F_x = 0$ at the anchoring point gives a pull

$$P = 2T \cos \alpha$$

on the leg. With the data given, the nurse increases the pull from $P_1 = 140$ N to $P_2 = 170$ N.

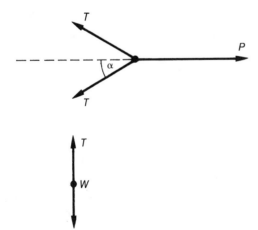

S13. Requiring $\Sigma M_O = 0$ for the pivot O (the elbow),

$$LW \cos \theta + \frac{L}{2} w \cos \theta - lF \cos \theta = 0,$$

so that $F = (L/l)W + (L/2l)w = 20W + 10w$. This greatly exceeds $W + w$ because the arm is (deliberately) an inefficient lever, as are most limbs. (An efficient lever would require large muscle contractions for small movements.)

S14. With the forces as shown in the Figure, the horizontal and vertical equilibrium conditions are

$$P\cos\theta - 2mg = 0 \tag{1}$$

and

$$P\sin\theta + N - mg = 0, \tag{2}$$

where N is the reaction force of the ground. If P is very slightly larger than the value specified by these conditions, the box will begin to move towards the first man. The condition specifying $\theta = \theta_c$ is $N = 0$, i.e. that the vertical component of the first man's pull would almost lift the box from the ground. Thus from (2), $P\sin\theta_c = mg$. Now eliminating P from (1), with $\theta = \theta_c$ we get $\tan\theta_c = 0.5$ or $\theta_c = 26.57°$. From (1) we get $P = 2mg/\cos\theta_c = 2.24mg$.

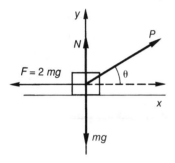

S15. As O is a fixed axis, we require $\Sigma M_O = 0$. The torques acting at O are the moments of the rod's weight and the string tension T. Since the weight acts through the midpoint of the rod, we must have

$$lT\sin\theta - \frac{l}{2}mg\cos\alpha = 0, \tag{1}$$

where l is the rod's length and θ is the angle of the string to the rod (see Figure). Note that we must use the force components acting perpendicular to the rod in taking moments, otherwise we will introduce the internal forces in the rod. Clearly $\theta = 90° - \alpha - \beta$, so (1) becomes $T\cos(\alpha + \beta) = (1/2)mg\cos\alpha$. From the vertical equilibrium of the hanging mass M we have $T = Mg$, so

$$M = m\frac{\cos\alpha}{2\cos(\alpha + \beta)} = m\frac{\cos 45°}{2\cos 60°} = 0.71m.$$

Let the reaction force P at the axis make an angle γ to the horizontal (see Figure). With the x and y axes horizontal and vertical the equilibrium conditions $\Sigma F_x = 0, \Sigma F_y = 0$ become

$$P \cos \gamma - T \sin \beta = 0,$$

$$P \sin \gamma + T \cos \beta - mg = 0.$$

With $T = Mg = 0.71mg$ and $\beta = 15°$, these are

$$P \cos \gamma = 0.71mg \times 0.26 = 0.185mg$$

$$P \sin \gamma = mg - 0.71mg \times 0.97 = 0.311mg.$$

Dividing the second equation by the first we get $\tan \gamma = 1.681$, so that $\gamma = 59.25°$. Then from the first equation we get $P = 0.185mg/\cos \gamma = 0.362mg$.

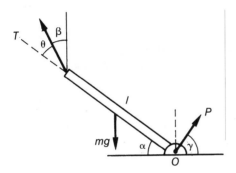

S16. Let the wire make an angle α to the horizontal (see Figure). Then requiring $\Sigma M_O = 0$ about O gives

$$\frac{l}{2} T \sin \alpha - lmg = 0.$$

Thus $T = 2mg/\sin \alpha$. Clearly $\sin \alpha = h/[h^2 + (l/2)^2]^{1/2}$. Writing $x = h/l$ we have

$$T = 2\left(1 + \frac{1}{4x^2}\right)mg.$$

When $T = T_{max} = 3mg$ we have $x^2 = 1/2$, so that $h_{min} = l/\sqrt{2} = 0.71l$.

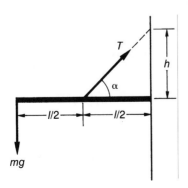

S17. Let the upper hinge be at A and the lower one at B and let the forces they exert on the door be F_A, F_B. The center of mass of the door is at its center O. Its weight acts vertically downwards through this point. Since hinge A carries all of this weight, F_B must be purely horizontal, while F_A must have both horizontal and vertical components (see Figure). Requiring $\Sigma M_A = 0$ gives

$$-\frac{w}{2} Mg + (h - 2d)F_B = 0.$$

With $d = w/4$ and $h = 3w$, we get $F_B = Mg/5$. The horizontal and vertical equilibrium conditions $\Sigma F_x = 0, \Sigma F_y = 0$ give

$$F_A \cos \alpha - F_B = 0,$$

$$F_A \sin \alpha - Mg = 0.$$

Thus rearranging and dividing these two equations gives $\tan \alpha = Mg/F_B = 5$. Hence $\alpha = 78.7°$. The last equation now gives $F_A = Mg/ \sin \alpha = 1.02 Mg$.

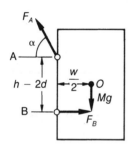

S18. The forces N_1, N_2 exerted by the wall and plane are normal to these two surfaces respectively (no friction). Thus N_1 is horizontal and N_2 makes an angle θ_2 to the vertical (see Figure). Then the horizontal and vertical equilibrium conditions $\Sigma F_x = 0, \Sigma F_y = 0$ imply

$$N_2 \cos \theta_2 = mg, \qquad (1)$$

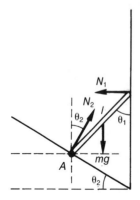

$$N_2 \sin \theta_2 = N_1. \tag{2}$$

Dividing (2) by (1) gives $\tan \theta_2 = N_1/mg$. Requiring $\Sigma M_A = 0$ about the point A where the rod touches the inclined plane gives

$$lN_1 \cos \theta_1 = \frac{l}{2} mg \sin \theta_1,$$

so that $\tan \theta_1 = 2N_1/mg$. Hence the required relation between the angles is $\tan \theta_1 = 2 \tan \theta_2$. With $\theta_2 = 30°$, this gives $\tan \theta_1 = 1.155$, so $\theta_1 = 49.1°$. From (1) we get $N_2 = mg/\cos \theta_2 = 1.15mg$, and substituting this into (2) gives $N_1 = N_2 \sin \theta_2 = 1.15mg \times (1/2) = 0.58mg$.

S19. Let N_1, N_2 be the normal reaction forces of the floor and wall, and f the frictional force exerted by the floor. Let the ladder have mass M and length L. Then the equilibrium conditions $\Sigma F_x = 0, \Sigma F_y = 0$ are

$$N_2 - f = 0, \tag{1}$$

$$N_1 - Mg = 0. \tag{2}$$

Requiring $\Sigma M_O = 0$ about the point O where the ladder is in contact with the floor (see Figure) gives

$$-LN_2 \sin \theta + \frac{L}{2} Mg \cos \theta = 0,$$

or

$$N_2 \tan \theta = \tfrac{1}{2} Mg. \tag{3}$$

Thus using (1) in (3) we get $f = Mg/2 \tan \theta$. Equilibrium is possible as long as f is no larger than the maximum possible frictional force, i.e. $f \leq N_1 \mu$. Now $N_1 \mu = Mg\mu$, using (2). Hence equilibrium requires $\tan \theta \geq 1/2\mu$, i.e. $\theta_m = \tan^{-1}(1/2\mu)$.

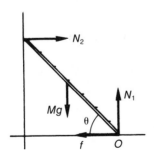

S20. The forces are as in the previous problem, with the addition of the worker's weight $2Mg$ acting at the top end of the ladder (see Figure). The equilibrium conditions $\Sigma F_x = 0, \Sigma F_y = 0$ thus become

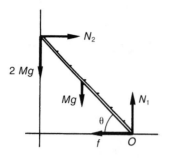

$$N_2 - f = 0,$$

$$N_1 - 2Mg - Mg = 0,$$

or $N_2 = f$, $N_1 = 3Mg$. Requiring $\Sigma M_O = 0$ about the contact point O now gives

$$-LN_2 \sin \theta + \frac{L}{2} Mg \cos \theta + L \times 2Mg \cos \theta = 0$$

or $N_2 \tan \theta = (5/2)Mg$. Thus $f = 5Mg/2 \tan \theta$. As before we require $f \leq N_1 \mu$ if the ladder is not to slip, which here becomes $f \leq 3Mg\mu$. Hence the condition determining θ_m is $\tan \theta \geq 5/6\mu$, i.e. $\theta_m = \tan^{-1}(5/6\mu)$, which is of course more restrictive than before.

S21. Let the mass of the platform be M, and let the load (of mass $M_L = 2M$) be at distance x from its left-hand edge. If the tensions in the two ropes are T_1, T_2, the equilibrium conditions $\Sigma F_x = 0, \Sigma F_y = 0$ become

$$T_2 \sin \theta_2 - T_1 \sin \theta_1 = 0, \tag{1}$$

$$T_1 \cos \theta_1 + T_2 \cos \theta_2 - 3Mg = 0. \tag{2}$$

(See Figure.)

Requiring $\Sigma M_O = 0$ about the position O of the load:

$$-xT_1 \cos \theta_1 - \left(\frac{L}{2} - x\right) Mg + (L - x)T_2 \cos \theta_2 = 0. \tag{3}$$

Substituting for the angles θ_1, θ_2 as given, and dividing (3) by $L/2$, equations (1–3) become

$$T_2\sqrt{3} = T_1, \tag{4}$$

$$T_1\sqrt{3} + T_2 = 6Mg, \tag{5}$$

$$\sqrt{3}\frac{x}{L}T_1 + \left(1 - 2\frac{x}{L}\right)Mg = \left(1 - \frac{x}{L}\right)T_2. \tag{6}$$

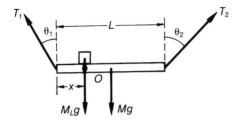

Solving (4, 5) for T_1, T_2 gives $T_1 = (3\sqrt{3}/2)Mg$, $T_2 = (3/2)Mg$. Substituting these values in (6) and dividing by Mg we get

$$\frac{9}{2}\frac{x}{L} + \left(1 - 2\frac{x}{L}\right) = \frac{3}{2}\left(1 - \frac{x}{L}\right),$$

with the solution $x = L/8$.

S22. If the cylinder is not to slide we require $\tan\theta \le \mu_s$ (see e.g. P7 above). It will overturn if and only if its center of gravity lies vertically outside the base, i.e. $\tan\theta > r/(h/2)$ (see Figure). Combining these two requirements shows that for $h > 2r/\tan\theta = 2r/\mu_s$ the cylinder will overturn. Note that this requirement is independent of θ.

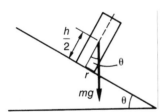

S23. The reaction force at the pivot will vanish if the two muscle pairs are arranged to be in vertical and horizontal equilibrium with the reaction force C acting downwards. $\Sigma F_x = 0$ requires

$$U\cos\theta_u - L\cos\theta_l = 0,$$

where U is the force exerted by the upper muscle pair. $\Sigma F_y = 0$ gives

$$U\sin\theta_u + L\sin\theta_l - C = 0.$$

Eliminating U between these two equations gives $C = L(\tan\theta_u\cos\theta_l + \sin\theta_l)$ $= 1.56L$ with the data given. This arrangement allows a larger biting or chewing force than would be exerted by either muscle group alone, and avoids creating large stresses on the jaw pivot.

S24. Horizontal equilibrium $\Sigma F_x = 0$ requires

$$F + F_2 - F_1 = 0. \tag{1}$$

Requiring $\Sigma M_O = 0$ about the root,

$$(l_1 + l_2)F - l_2 F_1 = 0. \tag{2}$$

From (2), $F_1 = (l_1 + l_2)F/l_2 = 0.35$ N. From (1), $F_2 = F_1 - F = 0.15$ N.

S25. If the pushing force is P and the player's mass is m, he will not overturn if the torque of P around his feet is smaller than that of his weight, i.e. we require

$$\left(\frac{5h}{8} + \frac{h}{4}\right) P \cos\theta < \frac{5h}{8} mg \sin\theta,$$

or $\tan\theta > 7P/5mg$. Horizontal equilibrium $\Sigma F_x = 0$ requires $P = f$, where $f \leq mg\mu$ is the frictional resistance at the player's feet. Thus the player begins to slide once P reaches the value $mg\mu$: he will not have overturned before this happens provided that $\tan\theta > 7\mu/5$. Hence the minimum angle of lean is $\theta_m = \tan^{-1}(7\mu/5)$.

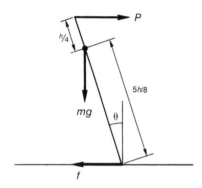

S26. Assume that the balance is level. Let the force exerted by the woman on the cord be F, and let the cord make an angle α to the vertical. Also, let the force exerted by the woman on the floor because of her weight be N'. Clearly $N' = N$, where N is the reaction force of the floor on the woman (see Figure). Requiring $\Sigma M_O = 0$ about the pivot O of the balance we have

$$lN + \frac{l}{2} F \cos\alpha = lMg$$

where l is the length of each arm of the balance. Canceling l,

$$N + \tfrac{1}{2} F \cos\alpha = Mg \tag{1}$$

(the weights of each side of the balance cancel). The vertical equilibrium condition $\Sigma F_y = 0$ for the woman is (see Figure)

$$N + F \cos\alpha = mg \tag{2}$$

as obviously $F' = F$. Eliminating $F \cos \alpha$ between (1) and (2) gives

$$N = (2M - m)g, \tag{3}$$

and thus from either (1) or (2) we get

$$F \cos \alpha = 2(m - M)g. \tag{4}$$

We require $N > 0$ if the woman is to remain on the platform, i.e. $2M > m$. Since she pulls the string we must have $F \cos \alpha > 0$, i.e. $m > M$. Combining these two requirements, the balance can remain level [for a suitable force F and angle α, cf. equation (4)] provided that m, M obey

$$M < m < 2M.$$

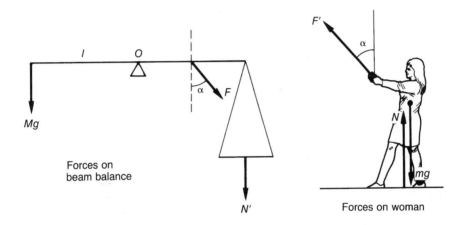

Forces on beam balance

Forces on woman

S27. If the lifting is slow, the situation is quasistatic. The pulley and mass are supported by *two* sections of rope, so $\Sigma F_y = 0$ gives $Mg = 2T$ or $T = Mg/2$. The woman only has to exert a force equal to one-half of the weight to be lifted. To lift the mass a height h, both the supporting section of rope must be shortened by an amount h. Thus the woman has to pull down a length $2h$ of rope.

S28. When the second pair of pulleys are added, the mass is supported by *four* sections of rope, so the vertical equilibrium condition $\Sigma F_y = 0$ becomes $4T = Mg$ or $T = Mg/4$. The four sections each have to be shortened by an amount h to raise the mass, so the woman now has to pull down a length $4h$ of rope.

S29. At the point A where the two levers touch, a torque G_2 on the left-hand shaft produces an upward force $F_2 = G_2/a$. To get the right-hand shaft just to turn requires $\Sigma F_y = 0$, i.e. F_2 must balance the resistive force $F_1 = G_1/b$. Thus the required torque is $G_2 = (a/b)G_1$. The calculation is precisely the same for the two gear wheels, as the teeth cause them to behave like a succession of levers, and steady motion implies that the forces are again in balance at A.

As the gear wheels cannot slip relative to each other, the upward velocities at A must be equal. If the right-hand wheel has angular velocity ω we have $a\Omega = b\omega$, so $\omega = (a/b)\Omega$.

The last three questions illustrate the principle of *gearing*: a smaller (larger) required force or torque corresponds to moving the load more slowly (rapidly).

S30. As the motion is quasistatic the forces and torques are effectively in equilibrium at all times. Simple geometry shows that the radius from the center of the cylinder to the contact point O makes an angle $\theta = 60°$ to the vertical (see Figure). When the rope is pulled horizontally, requiring $\Sigma M_O = 0$ about O gives

$$\frac{3}{2} R F_m = Rmg \sin \theta,$$

or $F_m = mg\sqrt{3}/3$.

If the reaction force of the curb is G and it makes an angle α to the horizontal, the equilibrium conditions $\Sigma F_x = 0$, $\Sigma F_y = 0$ are $F = G \cos \alpha$, $mg = G \sin \alpha$. Dividing these equations shows that $\tan \alpha = mg/F = \sqrt{3}$, or $\alpha = 60°$ for $F = F_m$ as above. As lifting proceeds, the lever arm of the rope pull F increases, while that of the weight decreases (see Figure), so we deduce that F_m decreases during lifting.

If the direction of the pull is allowed to vary, the best angle is obviously the one making the lever arm of the pull largest, i.e. perpendicular to the diameter passing through O. This is clearly at $60°$ to the horizontal (see Figure). Requiring $\Sigma M_O = 0$ about O now gives

$$2R F_m = Rmg \frac{\sqrt{3}}{2},$$

or $F_m = (\sqrt{3}/4)mg$.

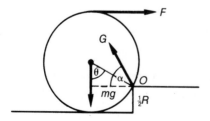

S31. (a) If $l < 2R$ the straw will slide until it reaches equilibrium, which by symmetry must occur when it is horizontal.

(b) If $l > 2R$ the horizontal equilibrium position is unattainable, and part of the straw will protrude from the glass (see Figure). Since the glass is smooth, the forces N_A, N_B exerted at the lower and upper contact points must be respectively perpendicular to the glass surface, i.e. directed towards the center of curvature O of the glass, and perpendicular to the straw (see Figure). Clearly the straw makes the same angle β with the horizontal and with N_A. Further the length AB is equal to $2R\cos\beta$. Now choosing the x-axis to lie along the straw and the y-axis perpendicular to it, the equilibrium conditions $\Sigma F_x = 0, \Sigma F_y = 0$ become

$$N_A \cos\beta = w\sin\beta, \tag{1}$$

$$N_A \sin\beta + N_B = w\cos\beta, \tag{2}$$

where w is the straw's weight. Requiring $\Sigma M_A = 0$ about A gives

$$2R\cos\beta N_B - \frac{l}{2}w\cos\beta = 0$$

or $N_B = (l/4R)w$. Substituting this into (2) gives

$$N_A \sin\beta = w(\cos\beta - \mu), \tag{3}$$

where we have written $\mu = l/4R$ for convenience. Thus dividing (1) by (3) gives

$$\frac{\cos\beta}{\sin\beta} = \frac{\sin\beta}{(\cos\beta - \mu)}.$$

Multiplying out, and using the identity $\sin^2\beta = 1 - \cos^2\beta$, we get a quadratic equation for $\cos\beta$:

$$2\cos^2\beta - \mu\cos\beta - 1 = 0,$$

with the solution

$$\cos\beta = \tfrac{1}{4}[\mu + \sqrt{\mu^2 + 8}],$$

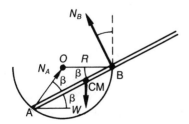

The other solution has $\cos\beta < 0$, which requires $\beta > 90°$, and is unphysical. Hence in the equilibrium position β is specified by

$$\cos\beta = \frac{l}{16R}\left\{1 + \left[1 + 128\left(\frac{R}{l}\right)^2\right]^{1/2}\right\},$$

and the length AB is $2R\cos\beta$, i.e.

$$AB = \frac{l}{8}\left\{1 + \left[1 + 128\left(\frac{R}{l}\right)^2\right]^{1/2}\right\},$$

which should be larger than $l/2$ if the straw is not to fall out of the glass.

S32. The woman lifts the mass slowly, so we can regard the situation as close to equilibrium. Using $\Sigma F_y = 0$ gives

$$Mg - R = 0,$$

where R is the tension in the rope, so $R = Mg$. Requiring $\Sigma M_E = 0$ about the elbow joint E,

$$Ta\sin(\theta + \phi) = fR\cos\phi.$$

Combining, we find

$$T = \frac{fR\cos\phi}{a\sin(\theta + \phi)} = \frac{8\cos\phi}{\sin(\theta + \phi)}Mg.$$

With $\theta = \phi$ we have $\sin(\theta + \phi) = \sin 2\theta = 2\sin\theta\cos\theta$, so $T \propto 1/\sin\theta$. The required tension in the biceps increases rapidly as the mass is raised and θ decreases.

S33. Let the tension in the rope be T. Using $\Sigma F_y = 0$ we get

$$T + Mg = F.$$

Taking moments about the point where the supports join the awning,

$$aT - (l - a)Mg = 0.$$

From the first equation $T = F - Mg$, so eliminating from the second gives $F = Mgl/a$. If instead two symmetrical sets of supports are used, $\Sigma F_y = 0$ immediately shows that $F = Mg/2$. With the data given, we get $F = 4900$ N in the first case and $F = 245$ N in the second.

☐ **KINEMATICS**

S34. The average speed is the total distance divided by the total time. The distance x_2 traveled after the stop is found from $x = vt$ as $x_2 = 90 \times 2 = 180$ km. Thus the total distance is $x = 50 + 180 = 230$ km. The total time includes the stop and is $t = 1/2 + 1/3 + 2 = 17/6$ h (20 min $= 1/3$ h). Hence the average speed is $v = x/t = 230/(17/6) = 81.2$ km/h.

S35. To answer the question we need to find the car's acceleration a. We must convert the car's velocity v to m s^{-1}. This gives $v = 100 \times 1000/3600 = 27.8$ m s^{-1}. Now using the kinematical formula $v = v_0 + at$ with $v_0 = 0, t = 10$ s and v as above, we find $a = v/t = 2.78$ m s^{-2}. The distance follows upon substituting these values into the formula $x = v_0 t + at^2/2$, giving $x = 2.78 \times 10^2/2 = 139$ m. The average velocity is this distance divided by the time 10 s, i.e. 13.9 m s^{-1}.

S36. The average velocity is $v_{\text{ave}} = s/t$, where t is the time to complete the journey. Clearly $t = s/2v_1 + s/2v_2 = s(v_1 + v_2)/2v_1 v_2$. Thus

$$v_{\text{ave}} = \frac{s}{t} = \frac{2v_1 v_2}{v_1 + v_2}.$$

This is always less than $v_{\text{mean}} = (v_1 + v_2)/2$ as the ratio is

$$\frac{v_{\text{ave}}}{v_{\text{mean}}} = \frac{4v_1 v_2}{(v_1 + v_2)^2}, \tag{1}$$

and since $(v_1 - v_2)^2 > 0$, we have $2v_1 v_2 < v_1^2 + v_2^2$, so $4v_1 v_2 < v_1^2 + 2v_1 v_2 + v_2^2 = (v_1 + v_2)^2$, so the rhs of (1) is always ≤ 1.

S37. The relative speed is $v_r = v_p - v_c = 60$ km/h. The officer has to travel $d = 0.5$ km relative to the car to catch it, so the time required is $t = d/v_r = 0.5/60$ h $= 30$ s.

S38. Concorde flies at speed v from East to West, relative to the Earth's atmosphere which turns with the Earth at speed $u = 2\pi R/d$ from West to East, where R is the Earth's radius and d is the length of the day. To make the Sun rise again requires $v > u = 2\pi \times 6400/24 = 1675$ km/h.

S39. We wish to use the formula $v^2 = v_0^2 + 2ax$; however, we must convert the velocity units first. Thus $v_0 = 100$ km/h $= 27.8$ m s^{-1}. Then with $a = -5$ m s^{-2} (deceleration $=$ negative acceleration) and $v = 0$ (the car comes to a stop) we find $x = -v_0^2/2a = 77.3$ m. If v_0 is increased by a factor 2, we see that x increases by a factor $2^2 = 4$. Thus the new stopping distance is 309 m.

S40. Using the kinematic formula $x = v_0 t + at^2/2$ with $v_0 = 0$ we find $a = 2x/t^2 = 2 \times 400/10^2 = 8$ m s^{-2}. Thus from $v = v_0 + at$ we get $v = 80$ m s^{-1}, or a speed of 288 km/h.

S41. In the kinematic formula $x = v_0 t + at^2/2$ we measure x upwards: we choose the roof level as $x = 0$, so the ground level is $x = -20$ m. Also $a = -g$. Then with $v_0 = 10$ m s^{-1} we get $-20 = 10t - 9.8t^2/2$, i.e.

$$4.9t^2 - 10t - 20 = 0.$$

The solution of this quadratic equation is $t = (10 \pm \sqrt{100 + 392})/9.8$. The negative root is not meaningful for this problem, so the answer required is the positive root $t = 3.28$ s. The impact velocity follows from the kinematic formula $v = v_0 + at$. With v_0, a as above and $t = 3.28$ s we find $v = -22.2 \, \mathrm{m\,s^{-1}}$, i.e. the ball hits the ground at $22.2 \, \mathrm{m\,s^{-1}}$ (the negative sign shows that the ball's motion is downwards).

S42. Choosing the positive x-direction downwards, we use the kinematic formula $x = v_0 t + at^2/2$. Here $a = g$ since the motion is downwards. In the first case we have $v_0 = 0$, thus $x = gt^2/2 = 9.8 \times 2^2/2 = 19.6$ m; this is the distance to the water surface. The impact velocity in this case is given by the formula $v = v_0 + at = gt = 9.8 \times 2 = 19.6 \, \mathrm{m\,s^{-1}}$. To find the initial velocity v_0 in the second case, we again use $x = v_0 t + at^2/2$, but now with x set equal to 19.6 m, $a = g$ and $t = 1$ s. This gives $19.6 = v_0 \times 1 + 9.8 \times 1^2/2 = v_0 + 4.9$. Thus $v_0 = 19.6 - 4.9 = 14.7 \, \mathrm{m\,s^{-1}}$. Here the impact velocity is given by $v = v_0 + at = 14.7 + 9.8 \times 1 = 24.5 \, \mathrm{m\,s^{-1}}$.

S43. The time needed for the car to overtake the truck is the time the truck takes to travel 32 m. From the kinematic formula $x = v_0 t + at^2/2$ with $v_0 = 0, a = a_2 = 1 \, \mathrm{m\,s^{-2}}$, we get $32 = t^2/2$ and thus $t = \sqrt{64} = 8$ s. The velocities of the car and truck follow from the formula $v = v_0 + at$, using the value of t above and $a = a_1$, $a = a_2$ respectively, with $v_0 = 0$ in both cases. We find $v_1 = 2 \times 8 = 16 \, \mathrm{m\,s^{-1}}$ and $v_2 = 1 \times 8 = 8 \, \mathrm{m\,s^{-1}}$. We can find the initial separation of the vehicles by subtracting 32 m (the distance traveled by the truck) from the distance x_1 traveled by the car by the time they are level. The latter is given by the formula $x = v_0 t + at^2/2$ with $v_0 = 0, a = a_2 = 2 \, \mathrm{m\,s^{-2}}$ and $t = 8$ s. This gives $x_1 = 2 \times 8^2/2 = 64$ m. Thus the initial separation was $\Delta x = 64 - 32 = 32$ m.

S44. From the kinematic formula $v^2 = v_0^2 + 2ay$ with $v_0 = 0$ (the rocket starts from rest), $y = 1000$ m and $v = 100 \, \mathrm{m\,s^{-1}}$ we find $a = v^2/2y = 10,000/2000 = 5 \, \mathrm{m\,s^{-2}}$. The time follows from $v = v_0 + at$ with $v_0 = 0$ as above and $a = 5 \, \mathrm{m\,s^{-1}}$ as deduced: this gives $t = v/a = 100/5 = 20$ s.

S45. The bullet reaches its maximum height when its vertical velocity $v = 0$. From the kinematic formula $v^2 = v_0^2 + 2ay$ with $v_0 = 30$ m/s, $a = -g$ we find a maximum height $y = v_0^2/2g = 900/(2 \times 9.8) = 45.92$ m. To find the velocity after $4\,$s we use the formula $v = v_0 + at$ with v_0, a as above, to find $v = 30 - 9.8 \times 4 = -9.2\,\mathrm{m\,s^{-1}}$. The negative sign shows that the bullet has passed its greatest height and is falling back. The corresponding height is given by the formula $y = v_0 t + at^2/2$ with v_0, a as above and $t = 4\,$s. We get $y = 30 \times 4 - 9.8 \times 16/2 = 120 - 78.4 = 41.6$ m.

S46. From the kinematic formula $v^2 = v_0^2 + 2ay$, with $v_0 = 0, a = g, y = h/2$ we find the velocity $v = \sqrt{gh}$ as the body starts the second half of its fall. Now using $y = v_0 t + at^2/2$ with $v_0 = v = \sqrt{gh}, a = g, y = h/2$ we find it falls the second half in a time t satisfying

$$\frac{h}{2} = \sqrt{gh}\,t + \frac{1}{2}gt^2.$$

Now we are told that $t = 1$ s, so

$$h = 2\sqrt{9.8h} + 9.8,$$

implying

$$(h - 9.8)^2 = 39.2h$$

or $h^2 - 58.8h + 96.04 = 0$. This quadratic equation has two roots, namely $h_1 = 57.1\,$m and $h_2 = 1.68\,$m. The latter solution is clearly impossible, as we know that the body falls for longer than 1 s, in which time it will have covered more than $gt^2/2 = 4.9$ m. Thus $h = 57.1\,$m.

S47. Using the kinematic formula $y = v_0 t + at^2/2$ with $y = H - h, v_0 = 0$, $a = g$ the man falls for a time t_m, where $H - h = gt_m^2/2$, i.e. $t_m = [2(H - h)/g]^{1/2} = 4.04$ s. Superwoman falls the same distance in time $t_m - 1 = 3.04$ s. Using the same kinematic formula again we have $H - h = v_0 t + gt^2/2$, or $80 = 3.04v_0 + 9.8 \times 3.04^2/2$, so that $v_0 = 11.4$ m s^{-1}.

S48. In the elevator frame the effective gravity is $g_{\mathrm{eff}} = g + a = 11.8$ m s^{-2}, and the ball simply rises and falls with respect to the elevator and boy under this acceleration. Using the kinematic formula

$$y = v_0 t - \frac{1}{2}g_{\mathrm{eff}}t^2$$

where y is the vertical distance from the boy's hand we see that $y = 0$ both at $t = 0$ and at $t = 2v_0/g_{\mathrm{eff}} = 0.85$ s.

S49. The time of flight is given simply by the vertical motion. This is governed by the equation $y = v_{0y}t - gt^2/2$. Here $v_{0y} = v_0 \sin\alpha$ with $v_0 = 300\,\mathrm{m\,s^{-1}}$ the muzzle velocity and $\alpha = 30°$ the elevation. The time of flight is given by setting $y = 0$, which gives $0 = v_{0y}t - gt^2/2$. The root $t = 0$ is trivial (the shell starts from $y = 0$ also), so we can divide through by t in this equation to get $t = 2v_{0y}/g = 300 \times 0.5/4.9 = 30.6\,\mathrm{s}$. The range follows from the horizontal motion, which is simply constant velocity at $v_{0x} = v_0 \cos 30° = 300 \times 0.866 = 260\,\mathrm{m\,s^{-1}}$. Thus the range is $x = v_{0x}t = 7956\,\mathrm{m}$.

S50. The maximum distance is achieved when the elevation angle is $45°$. We find the time of flight, as before, from the equation of vertical motion, finding $t = 2v_{0y}/g = 2 \times 25 \sin 45°/9.8 = 50 \times 0.71/9.8 = 3.62\,\mathrm{s}$. The best distance is thus $x = v_{0x}t = 25 \cos 45° \times 3.62 = 64\,\mathrm{m}$. To find the elevation of the faulty throw, we note that the range can be written quite generally as $x = v_{0x}t = v_{0x} \times 2v_{0y}/g = 2v_0^2 \sin\alpha\cos\alpha/g$. Using the trigonometric identity $\sin 2\alpha = 2\sin\alpha\cos\alpha$, this is $x = v_0^2 \sin 2\alpha/g$. With $x = 32\,\mathrm{m}$ for this throw and $v_0 = 25\,\mathrm{m\,s^{-1}}$ as before, we find $\sin 2\alpha = 0.5$. This has *two* solutions, $\alpha = 15°$ and $\alpha = 90° - 15° = 75°$. It is of course much more likely that the faulty throw was too flat than too steep, i.e. $\alpha = 15°$.

S51. We can rewrite the general range formula $x = v_0^2 \sin 2\alpha/g$ given in the last solution as $x = x_{\max} \sin 2\alpha$, where $x_{\max} = v_0^2/g$ is the maximum range. This shows that the maximum range is achieved when $\sin 2\alpha = 1$, i.e. $\alpha = 45°$, and that half the maximum range is achieved when $x_{\max} \sin 2\alpha = x_{\max}/2$, i.e. $\sin 2\alpha = 0.5$, so that $\alpha = 15°$ or $75°$ for half the range, independent of v_0.

S52. From the general range formula $x = 2v_0^2 \sin\alpha\cos\alpha/g$ used in the last two answers, we see that for given x and v_0 we have an equation for α, i.e. $\sin\alpha\cos\alpha = gx/2v_0^2$. If we find a solution $\alpha = \alpha_1$ of this equation, we can see that $\alpha_2 = 90° - \alpha_1$ is also a solution, since $\sin\alpha_1 = \cos\alpha_2$, $\cos\alpha_1 = \sin\alpha_2$. Clearly $\alpha_2 - 45° = 45° - \alpha_1$.

S53. (a) If the takeoff and landing points are at the same level we can use the range formula (see last three answers) in the form $\sin 2\alpha = xg/v_0^2$. With $x = 15$ m and $v_0 = 100$ km/h $= 27.8$ m s^{-1}, this gives $\sin 2\alpha = 0.19$, implying $\alpha = 5.5°$ (the alternative possibility $\alpha = 84.5°$ is rather unlikely!).
(b) If the bus takes off horizontally, the time of flight across the gap is $t = x/v_0$. Using the kinematic formula $y = v_{0y} + at^2/2$ during this time the bus falls a vertical distance $y = gt^2/2$, since it has zero vertical velocity initially. With the data given we find $y = gx^2/2v_0^2 = 1.4\,\mathrm{m}$.

S54. The time of flight follows from the horizontal motion as $t = x/v_0$, where v_0 is the muzzle velocity. The kinematic formula $y = v_{0y}t + at^2/2$ with $v_{0y} = 0$ shows that the bullet falls a distance $h = gt^2/2 = gx^2/2v_0^2$ below the horizontal. If the rifle is aimed correctly at some angle α to the horizontal, the range formula used in S50 above requires that $x = 2v_0^2 \sin\alpha\cos\alpha/g$, so $\sin\alpha\cos\alpha = xg/2v_0^2 = h/x$. For $h \ll x$, α is a small angle, so $\cos\alpha \approx 1$ and $\tan\alpha \approx \sin\alpha = h/x$. The rifleman should aim at a point $x\tan\alpha = h$ *above* the target.

S55. The time of flight is given by the vertical free-fall time from the airplane's height, with zero initial vertical velocity. Using $y = v_{0y}t - gt^2/2$ with $v_{0y} = 0$ and $y = -h$ we find $t = \sqrt{2h/g}$. Here h is the height, and y is negative because it is measured from the airplane's position. With $h = 2$ km $= 2000$ m we get $t = 20.2$ s. The tank's horizontal velocity is the same as that of the airplane and is thus $v_{0x} = 600$ km/h $= 167\,\mathrm{m\,s^{-1}}$. The horizontal distance traveled by the tank after release is thus $x = v_{0x}t = 167 \times 20.2 = 3370$ m. As the airplane and tank have exactly the same horizontal velocity, the airplane is always directly overhead the tank, including at the moment of impact.

S56. Since the bombs all have the same horizontal velocity as the bomber they lie on a vertical line directly underneath it at all times (see Figure). Each bomb takes exactly the same time to hit the ground, so they do so at intervals $\Delta t = 1$ s. Their release points differed by $v\Delta t = 194$ m, hence so do their impact points.

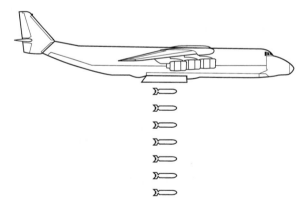

S57. The time of flight is given by the vertical motion as $t = 2v_{0y}/g$ (see S49). With $v_{0y} = v_0 \sin\alpha = 1000 \times 0.087 = 87\,\mathrm{m\,s^{-1}}$, we find $t = 2 \times 87/9.8 = 17.79$ s. The horizontal velocity of the shell *with respect to the ground* includes the tank's velocity u and is $v_{0x} = v_0\cos\alpha + u = 1000 \times 0.996 + 10 = 1006\,\mathrm{m\,s^{-1}}$. The range of the shell was therefore $x = v_{0x}t = 1006 \times 17.79 = 17{,}897$ m. During the shell's flight, the tank advanced a distance $ut = 10 \times 17.79 = 177.9$ m, so the separation of the tank and target at impact is the difference $17,897 - 177.9 = 17,719\,\text{m}$. The separation of the tank and target at the moment of firing is the shell's range *minus* the distance traveled by the target during the shell's flight, i.e. $17,897 - wt = 17,897 - 15 \times 17.79 = 17,630$ m.

S58. The horizontal distance traveled by the softball is $x = l + d = 38 + 2 = 40$ m. The time of flight is thus $t = x/v_{0x}$, where $v_{0x} = v_0 \cos \alpha = v_0/2$ ($\cos 60° = 0.5$) is the (constant) horizontal velocity of the ball, and v_0 is the unknown velocity of the throw. Hence $t = 2x/v_0 = 80/v_0$ s. Substituting this expression into the equation for vertical motion $y = v_{0y}t - gt^2/2$ with $v_{0y} = v_0 \sin \alpha = 0.866v_0$ and $y = h = 20$ m we find

$$20 = 0.866v_0 \times 80/v_0 - 9.8 \times (80/v_0)^2/2.$$

i.e. $20 = 69.3 - 31,360/v_0^2$, or $v_0 = \sqrt{31,360/49.3} = 25.2\,\text{m s}^{-1}$.

S59. Using the kinematic formulae, after time t we have horizontal and vertical displacements

$$x = ut, \tag{1}$$

$$y = vt - \frac{gt^2}{2}. \tag{2}$$

Using (1) to eliminate $t = x/u$, (2) becomes

$$y = \frac{v}{u}x - \frac{g}{2u^2}x^2. \tag{3}$$

This is a parabola. Clearly $y = 0$ at $x = 0$ and $x = r = 2uv/g$. The height h follows either directly by putting $x = r/2 = uv/g$ into the equation (3) of the parabola, giving $h = v^2/2g$, or by using the kinematic formula $v_y = v - gt$ for the vertical motion, which gives $t = v/g$ for the time at which the projectile reaches its greatest height ($v_y = 0$ there); giving t this value in (2) gives the same value for h.

S60. The athlete needs to launch the javelin at $45°$ to the ground (as viewed by a stationary observer) for maximum range (see S51). If she throws the javelin at angle θ in her own frame, she has to ensure that the horizontal and vertical components of its initial velocity seen by a stationary observer are equal, i.e.

$$v \sin \theta = v \cos \theta + \frac{v}{4},$$

where v is the speed of the throw. Thus $\sin \theta - \cos \theta = 0.25$, which is satisfied for $\theta = 55°$.

S61. Equations (1–3) of S59 hold here too. As the pea is aimed directly at the cat, the boy chose the velocity components u, v so that the straight line $y/x = v/u$ passes through the cat's initial position. Equation (2) shows that at time t the pea is a distance $gt^2/2$ below this line. But the cat falls from rest on this line, so at time t it too is a distance $gt^2/2$ below this line. Thus when the pea reaches the line of the cat's fall it will have the same vertical displacement from the line, i.e. it hits the cat.

S62. The skier takes off from the top of the hump with horizontal and vertical velocity components $u, 0$. From equations (1, 2) of Solution 61 we have horizontal and vertical displacements $x = ut, y = -gt^2/2$ at time t. This gives the dashed trajectory in the Figure. The skier lands when $y = -x \tan \alpha$, or $-gt^2/2 = -ut \tan \alpha$, i.e. $t = (2u/g) \tan \alpha$. Using $u = 100$ km/h $= 27.8$ m s^{-1}, we find $t = 56.7 \tan 25° = 2.65$ s.

The vertical velocity $v = 5$ m s^{-1} allows the skier instead to "pre-jump" the crest of the hump, i.e. take the trajectory indicated in the Figure by the dotted curve, since $h < v^2/2g$. If executed perfectly, this trajectory would have takeoff speed given by $w = (2gh)^{1/2} < v$ and take a time $t_{\text{pre}} = w/g = (2h/g)^{1/2}$. With the data given $t_{\text{pre}} = (2/9.8)^{1/2} = 0.45$ s. The pre-jump trajectory saves more than 2 s of time in the air. The speed difference between skiing on snow and airborne implies a significant overall time saving, and pre-jumping is a standard competition technique.

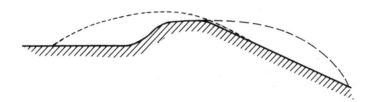

S63. The man should arrange that his velocity with respect to the river banks points directly towards his girlfriend. Thus he should swim at angle α to the shortest distance across the river, partly into the current so that he cancels it, i.e. $v_s \sin \alpha = v_w$ (see Figure). Thus $\sin \alpha = v_w/v_s = 0.5$, or $\alpha = 30°$. His

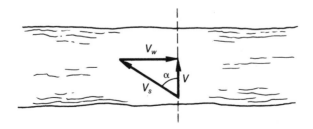

velocity across the river is then $v_s \cos \alpha = 0.87\,\mathrm{m\,s^{-1}}$, so the crossing takes a time $t = L/0.87 = 115$ s $= 1.9$ min.

S64. We have the two velocity triangles shown in the Figure. Here v_{pA}, v_{pB} are the airplane's speed relative to trains A and B respectively. From that for A we have that he sees the airplane's speed as $v_{pA} = v \tan \alpha$. Using this in the triangle for B we have $v \tan \alpha = 2v \tan \theta$, which gives $\tan \alpha = 2 \tan 30° = 1.155$ and thus $\alpha = 49°$. From the triangle for A we have $v_g = v/\cos \alpha = 60/\cos 49° = 91.5$ km/h.

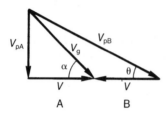

S65. In the runner's reference frame the rain has a horizontal velocity component exactly equal and opposite to the runner's velocity (see Figure). The total velocity of the rain is then $(u^2 + v^2)^{1/2}$, at angle $\phi = \tan^{-1} v/u$ to the vertical. If the runner leans forward at angle θ he presents total effective area $A = A_f \sin(\phi - \theta) + A_t \cos(\phi - \theta)$ to the rain (see Figure). If $A_t < A_f$ this is obviously smallest when $\theta = \phi$, so that $A = A_t$, i.e. all the rain falls on the runner's head and shoulders. As it falls with velocity $(u^2 + v^2)^{1/2}$ and effective density ρ, the total mass of water absorbed in unit time is

$A_t\rho(u^2 + v^2)^{1/2}$. The runner spends a time l/v in the rain, so the minimum amount of water he absorbs is

$$m = A_t l \rho \frac{(u^2 + v^2)^{1/2}}{v}.$$

Thus even if the runner could run much faster than the rain falls ($v \gg u$) he would still absorb at least a mass $A_t l \rho$ of water (actually much more, as he cannot lean forward at an angle $\theta = \phi = \tan^{-1} v/u \approx 90°$!). In practice $v \ll u$, and $m \approx A_t l \rho u/v$. This gives the answer to the often-asked question as to whether running faster in rain merely gets the runner wet faster – on the contrary, doubling the speed v actually halves the mass of water absorbed.

S66. The main problem in believing the man's claim are the accelerations required to reduce the relative speed of the two cars to 10 km/h or less. If the second car did not manage to turn and accelerate significantly, the first car must have braked hard enough to reduce its speed from $v_0 = 70$ km/h $= 19.44$ m s^{-1} to $v = 10$ km/h $= 2.77$ m s^{-1} in a distance $x = 4$ m. Using the kinematic formula $v^2 = v_0^2 + 2ax$ we find an acceleration $a = -46$ m s^{-2}, or $a \approx -4.7g$. This is far more rapid braking than is likely (typically $|a| < g$) even allowing for the first driver's reaction time. If instead the second car managed to turn and accelerate to 60 km/h in 4 m, the same formula requires the car to have an acceleration $a \approx 3.5g$. This is again implausibly high. Obviously one can imagine a combination of these two possibilities in which the first car slowed somewhat and the second accelerated by some amount. However, in all cases the required accelerations are too large to be believable.

☐ NEWTON'S SECOND LAW

S67. To find how the masses move we need their accelerations. In this problem they have the same value a because the string is under tension. The only force acting in the direction of motion on the mass m_1 is the string tension T (see Figure), so the equation of motion of m_1 is

$$m_1 a = T.$$

The forces acting on mass m_2 are T and its own weight $m_2 g$ (see Figure), so

$$m_2 a = m_2 g - T.$$

Adding these two equations eliminates T, i.e.

$$(m_1 + m_2)a = m_2 g,$$

so

$$a = \frac{m_2}{m_1 + m_2} g.$$

With the masses given we find $a = 0.097 \, \text{m s}^{-2}$. From the kinematic formula $x = v_0 t + at^2/2$ with $v_0 = 0, t = 10 \, \text{s}$, we get the distance traveled in the first 10 s as $x = 4.85 \, \text{m}$. The same formula gives the time to travel a distance $x = 1 \, \text{m}$ from rest as $t = \sqrt{2x/a}$. Here this is $4.54 \, \text{s}$.

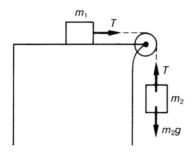

S68. The resultant upward force on the mass is $\Sigma F_y = T - mg$, where T is the tension. From Newton's second law we have $a = \Sigma F_y/m = T/m - g$. The maximum acceleration follows upon substituting the maximum allowed tension $T = 500 \, \text{N}$, giving $a_{\text{max}} = 500/20 - 9.8 = 15.2 \, \text{m s}^{-2}$. Using this acceleration in the kinematic formula $y = v_0 t + at^2/2$ with $v_0 = 0$ and $t = 2 \, \text{s}$ we get $y = 15.2 \times 2^2/2 = 30.4 \, \text{m}$ for the distance the mass has traveled.

S69. The motion up the inclined plane is one-dimensional, and we define the distance from the initial position to be x. To use the kinematic formulae we first need the acceleration. The resultant force component on the body in the x-direction is $\Sigma F_x = -mg \sin \alpha$ (see Figure). (The resultant force normal to the plane is zero as the component of weight in this direction is balanced by the normal reaction force of the plane.) Thus the acceleration is $a = \Sigma F_x/m = -g \sin \alpha$. In this case $\alpha = 30°$ and thus $a = -g \times 0.5 = -4.9 \, \text{m s}^{-2}$. The kinematic formula to use here is $v = v_0 + at$. With $v = 0$ (the turning point) and $v_0 = 5 \, \text{m s}^{-1}$ we get $t = -v_0/a = 5/4.9 = 1.02 \, \text{s}$.

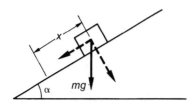

S70. We take the motion of the lighter body to define the positive x-direction. (The heavier body moves downwards.) Considering each body separately, we can use Newton's second law and the resultant forces on them to write

$$m_1 a = \Sigma F_x(1) = T - m_1 g,$$

$$m_2 a = \Sigma F_x(2) = m_2 g - T,$$

where a is the common acceleration of the two masses (see Figure). Adding the two equations we find $(m_1 + m_2)a = (m_2 - m_1)g$, and thus

$$a = \frac{m_2 - m_1}{m_1 + m_2} g,$$

giving $a = 5 \times 9.8/15 = 3.27\,\mathrm{m\,s}^{-2}$. From either of the equations we can now find T by substituting for a. From the first equation we find $T = m_1 (a + g) = 5 \times (3.27 + 9.8) = 65.35$ N.

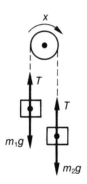

S71. Let the angle we seek be θ and the tension be T. The resultant forces on the mass in the x and y directions are then (see Figure) $\Sigma F_x = T \sin \theta$, $\Sigma F_y = T \cos \theta - mg$. There is no vertical motion, so $\Sigma F_y = 0$, but in the horizontal direction, Newton's second law requires $\Sigma F_x = ma$. Thus

$$T \cos \theta - mg = 0$$

$$T \sin \theta = ma.$$

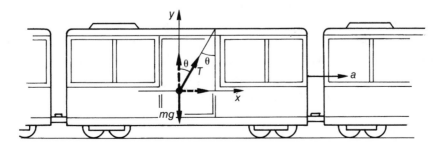

Putting $a = 0.1g$ in the second equation and rearranging the first we get $T \cos \theta = mg, T \sin \theta = 0.1mg$. Dividing the second equation by the first we get $\tan \theta = 0.1$, with the solution $\theta = 5.7°$. Using this value in the first equation we find $T = mg/\cos 5.7° = 1.005mg$. Note that the tension is larger than the weight, because the subway car accelerates the mass through the tension in the string.

S72. In the vertical direction, the forces acting on the person are $\Sigma F_y = N - mg$, where N is the normal force exerted by the elevator floor. By Newton's second law, $\Sigma F_y = ma$, so $N = m(a + g) = 1.1mg$. (Note: this is the person's effective weight.) The vertical force on the elevator and its contents is $\Sigma F_y = T - Mg - mg$. By Newton's second law this is equal to $(M + m)a = 0.1(M + m)g$. Thus $T - (M + m)g = 0.1(M + m)g$, so $T = 1.1(M + m)g$.

S73. The motion of each mass is one-dimensional, and they must move equal amounts along the wedge faces. The resultant forces on the masses along the wedge faces to the left can be written as

$$\Sigma F_1 = mg \sin \theta_1 - T$$

$$\Sigma F_2 = T - Mg \sin \theta_2.$$

If the masses are to remain stationary, both resultant forces must vanish. With $\sin 53° = 0.8, \sin 37° = 0.6$ this gives $0.8mg - T = 0, T - 0.6Mg = 0$. Eliminating T between these equations gives $0.8mg = 0.6Mg$, so $M/m = 0.8/0.6 = 1.33$. The tension T follows from the first relation as $T = 0.8mg$.

S74. After the additional mass m has been added, the resultant forces on each mass are $\Sigma F_1 = (M + m)g - T, \Sigma F_2 = T - Mg$. Each mass has the same acceleration a, which by Newton's second law obeys $\Sigma F_1 = (M + m)a, \Sigma F_2 = Ma$. Substituting these expressions into the first pair of equations gives

$$(M + m)g - T = (M + m)a,$$

$$T - Mg = Ma.$$

Adding these equations eliminates T, and we get $(M + m)g - Mg = (M + m)a + Ma$, so $mg = (2M + m)a$ or $a = mg/(2M + m) = 0.01Mg/(2M + 0.01M) = 4.98 \times 10^{-3}g$. After the extra mass is removed, the masses move with a constant velocity (the forces balance) whose value is $v = H/t = 0.312/1 = 0.312 \, \text{m s}^{-1}$. This is also the velocity acquired after accelerating under the extra weight. Using the formula $v^2 = v_0^2 + 2ax$ with $v_0 = 0, x = h = 1$ m, and a as above, we get $v^2 = 2ah = 2 \times 4.98 \times 10^{-3}g \times 1 = 9.96 \times 10^{-3}g$. Using $v = 0.312 \, \text{m s}^{-1}$ as found above gives $g = 0.312^2/9.96 \times 10^{-3} = 9.77\,\mathrm{m\,s^{-2}}$. (The deviation from the best value $g = 9.81\,\mathrm{m\,s^{-2}}$ is a result of experimental error.)

S75. The bullet's time of flight is equal to the free-fall time from rest at height h, since the bullet had zero initial vertical velocity. Thus using $y = v_0 t - gt^2/2$ with $v_0 = 0, y = -h$ we find $t = \sqrt{2h/g} = \sqrt{2 \times 1.5/9.8} = 0.553\,\mathrm{s}$. The horizontal range s and the time t give the muzzle velocity from the relation $x = vt$ with $x = s = 500$ m and $t = 0.553$ s. Hence $v = s/t = 904\,\mathrm{m\,s^{-1}}$. To find the force on the bullet we need the acceleration it experiences inside the gun. This is given by the formula $v^2 = v_0^2 + 2ax$ with $v_0 = 0$ (the bullet accelerates from rest) and $x = l = 0.5$ m. Thus $a = v^2/2l = 904^2/2 \times 0.5 = 8.17 \times 10^5\,\mathrm{m\,s^{-2}}$. The force on the bullet is (by Newton's second law) $F = ma$, where $m = 0.01$ kg, so that $F = 8170$ N.

S76. We choose the downward direction of motion as positive. We can find the acceleration from the kinematic formula $v = v_0 + at$ with $v_0 = 20\,\mathrm{m\,s^{-1}}$, $v = 5$ m s^{-1} and $t = 5$ s. Thus $a = (v - v_0)/t = (5 - 20)/5 = -15/5 = -3\,\mathrm{m\,s^{-2}}$. The minus sign shows that the skydiver decelerates. The forces acting on the skydiver during deceleration are her weight mg downwards and the tension T in the parachute cords upwards. Hence the resultant downward force is $\Sigma F_y = mg - T$. Using Newton's second law this is equal to ma, so $mg - T = ma$. Hence $T = m(g - a) = 50(9.8 - (-3)) = 50 \times 12.8 = 640\,\mathrm{N}$. The resultant force on the skydiver is $\Sigma F_y = mg - T = 50 \times 9.8 - 640 = -150$ N. Note that this is equal to ma, as it must be according to Newton's second law. The force acts upwards, as the skydiver's downward motion is decelerated.

S77. During braking the resultant horizontal force on the car is

$$\Sigma F_x = -f,$$

where $f = \mu N$ is the frictional force and we have chosen the x-direction to lie in the direction of motion. Here N is the normal force exerted by the road on the car tires. The vertical resultant force on the car vanishes, i.e.

$$\Sigma F_y = N - mg = 0,$$

so that $f = \mu N = \mu mg$. Newton's second law for the horizontal motion gives $ma = \Sigma F_x = -f = -\mu mg$. Thus $a = -\mu g$. The negative sign implies deceleration.

Using the kinematic formula $v = v_0 + at$ with $v = 0$ (complete stop) and a as above, we get the stopping time $t = -v_0/a = v_0/(\mu g)$. Since $v_0 = 60\,\mathrm{km/h} = 16.67\,\mathrm{m\,s^{-1}}$ this gives the stopping time $t = 16.67/(0.5 \times 9.8) = 3.4$ s. The stopping distance follows from the formula $x = v_0 t + at^2/2 = 16.67 \times 3.4 - 0.5 \times 9.8 \times 3.4^2/2 = 28.4\,\mathrm{m}$.

S78. The horizontal and vertical forces acting on the sled (see Figure) give the resultant forces $\Sigma F_x = F - f$, $\Sigma F_y = N - mg$ where f is the frictional force and N is the normal force exerted by the snow. For constant velocity both resultant forces must vanish, so that $F = f$ and $N = mg$. The frictional force is given by $f = \mu N$, so using the value for N we find $f = \mu mg$; thus $F = f = \mu mg = 0.1 \times 10 \times 9.8 = 9.8\,\text{N}$.

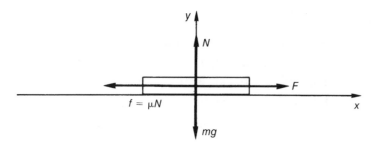

S79. Choosing the x-coordinate to run downwards along the slope and the y-coordinate as its upward normal, the resultant forces on the static skier are (see Figure) $\Sigma F_x = mg\sin\alpha - f$, $\Sigma F_y = N - mg\cos\alpha$. Both resultant forces vanish, so that

$$N = mg\cos\alpha,$$

$$f = mg\sin\alpha.$$

Until the skier begins to move, f is *smaller* than $\mu_s N$; the motion starts when $f = \mu_s N$. Substituting this into the second equation and dividing it by the first, we find $\mu_s = \tan\alpha = \tan 15° = 0.268$ (cf. P7). After the motion starts, the coefficient of friction drops to a value $\mu = 0.1$, and $f = \mu N$ always holds. Now, ΣF_x has the nonzero value $mg\sin\alpha - \mu N$, where N is the

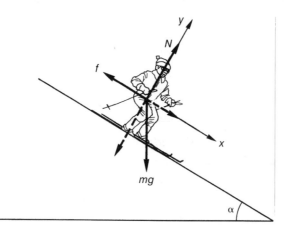

same as before. Replacing N in the last expression, we get $\Sigma F_x = mg \sin \alpha - \mu mg \cos \alpha$. Newton's second law now gives the acceleration

$$a = \Sigma F_x / m = g(\sin \alpha - \mu \cos \alpha) \text{ or}$$

$$a = 9.8 \times (\sin 15° - 0.1 \cos 15°) = 1.59 \, \text{m s}^{-2}.$$

The velocity v and distance x after 5 s follow from the kinematic formulae $v = v_0 + at$, $x = v_0 t + at^2/2$ respectively. With $v_0 = 0$ and a as found above these give $v = 1.59 \times 5 = 7.95 \, \text{m s}^{-1}$ and $x = 1.59 \times 5^2/2 = 19.9$ m.

S80. Using the Figure, we see that

$$\Sigma F_x = F \cos \alpha - f,$$

$$\Sigma F_y = F \sin \alpha + N - Mg,$$

where f is the frictional force given by $f = \mu N$, with N the normal force on the timber. Using Newton's second law, $\Sigma F_x = Ma$, where a is the acceleration, and $\Sigma F_y = 0$. (The rope does not lift the timber completely off the ground: if it did, N would become formally negative.) Substituting these three relations into the pair of equations above, we get

$$Ma = F \cos \alpha - \mu N,$$

$$0 = F \sin \alpha + N - Mg.$$

From the second equation, $N = Mg - F \sin \alpha$. Putting this into the first equation gives

$$Ma = F \cos \alpha - \mu(Mg - F \sin \alpha) = F(\cos \alpha + \mu \sin \alpha) - \mu Mg.$$

Substituting the numerical values given we get $a = 3 \times 0.97 - 0.2 \times 1 \times 9.8$, i.e. $a = 0.95 \, \text{m s}^{-2}$. From the equation for N we find

$$N = 100 \times 9.8 - 300 \times 0.5 = 830 \, \text{N}.$$

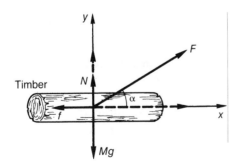

Note that this is positive, but smaller than the weight Mg of the timber, as the dragging force has an upward component.

S81. We take the x-direction to run up the slope, and the y-direction normal to the slope. The resultant forces on the body in its upward motion are (see Figure 1)

$$\Sigma F_x = -mg \sin \alpha - f,$$

$$\Sigma F_y = N - mg \cos \alpha.$$

With $\Sigma F_y = 0$ and $f = \mu N$ as usual, we get (eliminating N) $\Sigma F_x = -mg \sin \alpha - \mu mg \cos \alpha$. The acceleration a_1 follows from Newton's second law i.e. $a_1 = \Sigma F_x/m = -g(\sin \alpha + \mu \cos \alpha)$ i.e. $a_1 = -9.8(0.342 + 0.2 \times 0.940) = -5.19 \,\mathrm{m\,s^{-2}}$. The negative sign implies that this is downwards. The time t_1 is given by the formula $v = v_0 + at$ with $v = 0$ (turning point), $v_0 = 10 \,\mathrm{m\,s^{-1}}$ and $a = a_1$ as above. We find $t_{\mathrm{up}} = -v_0/a_1 = -10/(-5.19) = 1.93 \,\mathrm{s}$. The distance s can be found from $s = x = v_0 t + at^2/2$ with $v_0 = 10 \,\mathrm{m\,s^{-1}}$, $a = a_1 = -5.19 \,\mathrm{m\,s^{-2}}$ and $t = t_{\mathrm{up}} = 1.93 \,\mathrm{s}$. We find $s = 10 \times 1.93 - 5.19 \times (1.93)^2/2 = 9.63$ m.

In the downward motion, the resultant force in the y-direction is the same, but the frictional force f is *reversed* in the formula for ΣF_x, because friction always opposes the motion (see Figure 2). This gives $\Sigma F_x = -mg \sin \alpha + \mu mg \cos \alpha$, and thus the acceleration $a_2 = -g(\sin \alpha - \mu \cos \alpha)$ in the downward motion. Hence $a_2 = -9.8(0.342 - 0.2 \times 0.940) = -1.51 \,\mathrm{m\,s^{-2}}$. The time t_{down} follows from the formula $x = v_0 t + at^2/2$ with $v_0 = 0$ (turning point), $x = -s$ (the motion is downwards, i.e. to negative x) and $a = a_2$. Thus $t_{\mathrm{down}} = \sqrt{2(-s)/a_2} = \sqrt{2 \times 9.63/1.51} = 3.57 \,\mathrm{s}$.

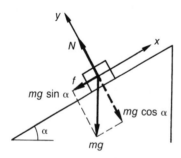

Fig 1 Upward motion

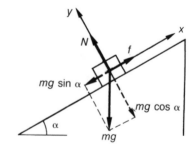

Fig 2 Downward motion

S82. Let the tension in the string be T. If T is too large the mass m moves upwards. The maximum allowed value follows from $T_1 = mg \sin \alpha + f$, where $f = \mu_s N = \mu_s mg$ is the frictional force (see Figure). Thus $T_1 = mg(\sin \alpha + \mu_s \cos \alpha)$.

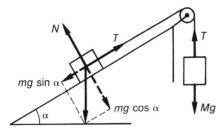

$f = \mu_s N$ acts to the left or to the right (opposite to the direction of the force trying to move the mass).

If T is too small the mass m moves downwards. The minimum allowed value is $T_2 = mg(\sin\alpha - \mu_s\cos\alpha)$ (the frictional force is reversed compared with that in T_1). Equilibrium requires that T lie between these two values T_1, T_2, i.e. $T_2 < T < T_1$. Thus

$$mg(\sin\alpha - \mu_s\cos\alpha) < T < mg(\sin\alpha + \mu_s\cos\alpha).$$

But we also know that $T = Mg$, since the hanging mass is at rest in equilibrium. Thus inserting this in the last inequality gives

$$m(\sin\alpha - \mu_s\cos\alpha) < M < m(\sin\alpha + \mu_s\cos\alpha).$$

Since $\sin\alpha = 0.6, \cos\alpha = \sqrt{1 - \sin^2\alpha} = 0.8$, so M lies between $M_1 = (0.6 - 0.2 \times 0.8)m = 0.44m$ and $M_2 = (0.6 + 0.2 \times 0.8)m = 0.76m$, or

$$0.44 < \frac{M}{m} < 0.76.$$

S83. The chain is uniform, so the weights w_1, w_2 of its vertical and horizontal parts are $w_1 = l_1 w/l$, $w_2 = (l - l_1)w/l$ respectively, where w is its total weight. The chain tension T at the edge of the table must balance the weight of the hanging part, i.e. $T = w_1$. The *maximum* hanging length l_1 is determined by the condition that the horizontal frictional force $f = \mu_s w_2$ can just balance this tension. Thus $\mu_s w_2 = f = T = w_1$, so substituting for w_1, w_2, we find $\mu_s(l - l_1) = l_1$, i.e.

$$l_1 = \frac{\mu_s}{1 + \mu_s}l.$$

As expected, for the largest possible friction coefficient $\mu_s = 1$, exactly one half of the chain can overhang the table.

S84. As the masses are equal, the direction of motion will be down the steeper slope. The resultant force on the ascending mass m in the direction of motion is

$$\Sigma F_1 = T - mg\sin\theta_2 - \mu mg\cos\theta_2.$$

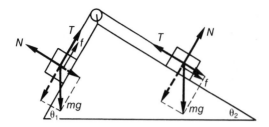

The resultant force on the other mass m is

$$\Sigma F_2 = mg \sin \theta_1 - T - \mu mg \cos \theta_1.$$

Motion at constant velocity implies that both forces vanish. Adding the two equations with $\Sigma F_1 = \Sigma F_2 = 0$, we get $0 = mg(\sin \theta_1 - \sin \theta_2) - \mu mg(\cos \theta_1 + \cos \theta_2)$. Thus $\mu = (0.8 - 0.6)/(0.8 + 0.6) = 0.2/1.4 = 0.143$.

S85. We take the x-direction up the slope and the y-direction normal to it. At the moment when the mass begins to move, $F = F_{\max}$ and the resultant forces (see Figure) are

$$\Sigma F_x = F_{\max} \cos \alpha - mg \sin \alpha - \mu_s N,$$

$$\Sigma F_y = N - F_{\max} \sin \alpha - mg \cos \alpha.$$

Both forces must vanish, so we can use the second equation to write

$$N = F_{\max} \sin \alpha + mg \cos \alpha$$

and thus

$$F_{\max} = mg \frac{\sin \alpha + \mu_s \cos \alpha}{\cos \alpha - \mu_s \sin \alpha}.$$

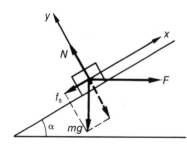

S86. As long as the box remains stationary on the accelerating truck their accelerations are the same. The only horizontal force acting on the box is friction (see Figure). Hence in this case we must have $f = ma$, where f is the frictional force, m is the mass of the box and a is the acceleration. Since f has a maximum value $f_{\max} = \mu_s N = \mu_s mg$, we obtain the maximum allowed acceleration of the truck as $a_t = \mu_s g = 0.3 \times 9.8 = 2.94 \, \text{m s}^{-2}$.

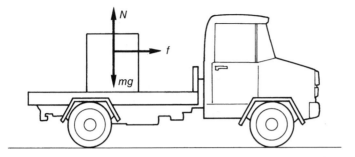

If this is even slightly exceeded, the coefficient of friction drops to $\mu = 0.2$. Again the only horizontal force on the box is friction, but now this is too small to prevent the box sliding backwards *with respect to the truck*. We now have $f = ma_b$, where f is now the sliding frictional force $f = \mu N = \mu mg$ and a_b is the acceleration of the box *with respect to the ground*. Thus $a_b = f/m = \mu g = 0.2g = 1.96\,\mathrm{m\,s}^{-2}$.

The distances x_t, x_b traveled by the truck and the box relative to the ground in the first second are given by the formula $x = v_0 t + at^2/2$. The initial velocity v_0 with respect to the ground is the same for the truck and the box, so

$$x_t = v_0 t + \frac{a_t t^2}{2},$$

$$x_b = v_0 t + \frac{a_b t^2}{2}.$$

The distance traveled by the box with respect to the truck is $\Delta x = x_b - x_t$. Subtracting the first equation from the second we get $\Delta x = (a_b - a_t)t^2/2$. Note that $a_b - a_a$ is the acceleration of the box with respect to the truck. This gives $\Delta x = (1.96 - 2.94)/2 = -0.49$ m. Thus the box slides 0.49 m backwards on the truck in time $t = 1$ s.

S87. Clearly the monitor cannot move with respect to the computer without also moving with respect to the table. The condition that the monitor should not move *with respect to the table* is found from the balance of horizontal forces on the monitor. This gives $F - f_1 = 0$ (see Figure). Since f_1 has a maximum value of μmg, this gives $F_{\max} = \mu mg$. We can now show that the full monitor–computer system does not move in this case. The external horizontal force acting on this system is $\Sigma F_x = F - f_2$, where f_2 is the frictional force between the computer and the table. For the case $F = F_{\max} = \mu mg$, we see that this is less than the maximum allowed value $3\mu mg$ of f_2, so the system remains at rest. Hence the monitor does not move with respect to the computer either.

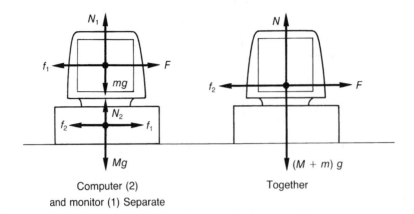

Computer (2) and monitor (1) Separate

Together

If $F = 2F_{\text{max}} = 2\mu mg$, the resultant horizontal forces are for the monitor

$$\Sigma F_1 = F - f_1 = 2\mu mg - \mu mg = \mu mg,$$

and for the computer

$$\Sigma F_2 = f_1 - f_2 = \mu mg - f_2.$$

Since the maximum allowed value of f_2 is $3\mu mg$ as before, the second equation implies that the computer does not move (friction equal to applied force up to a maximum). The monitor, however, moves with respect to the table (hence with respect to the computer) with acceleration given by Newton's second law applied to the first equation: $\mu mg = ma$, so $a = \mu g$.

S88. The condition that the whole book–paper system should move at all is found by comparing the external applied force P with the maximum static frictional force between the system and the table, i.e. $f_2 = \mu(M + m)g = 0.11Mg$. Thus P must exceed $0.11Mg$ for any motion to occur. Assuming that this condition holds so that the book and paper both move, we consider the case where there is no relative motion of the book and paper. The resultant horizontal forces on the book and paper are then respectively

$$\Sigma F_1 = f_1,$$

$$\Sigma F_2 = P - f_1 - f_2 = P - f_1 - \mu(M + m)g$$

where f_1 is the frictional force between the book and the paper (see Figure) and f_2 is the force between the paper and the table, i.e. $f_2 = \mu(M + m)g$. Let the common acceleration of the book and paper be a. Using Newton's second law in both equations gives

$$f_1 = Ma,$$

$$P - f_1 - \mu(M + m)g = ma.$$

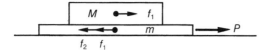

Adding the two equations and solving for P gives

$$P = (M + m)(a + \mu g).$$

Now f_1 has a maximum value $f_1(\text{max}) = \mu Mg$, so since $f_1 = Ma$, a also has a maximum value $a_{\text{max}} = f_1(\text{max})/M = \mu g$. Above this value the book cannot accelerate as fast as the paper, which can therefore be extracted. Substituting $a = a_{\text{max}}$ into the equation for P above gives $P = (M + m)(\mu g + \mu g) = 2\mu(M + m)g = 0.22Mg$. Thus $P_{\text{extract}} \geq 0.22Mg$.

☐ WORK, ENERGY AND POWER

S89. Only the horizontal component of the force F does work (there is no motion in the vertical direction). The horizontal component is $F_x = F \cos\theta = 5 \times 0.984 = 4.92\,\text{N}$. To find the work done we need the distance traveled in 5 s. Newton's second law gives the horizontal acceleration as $a = F_x/m = 4.92/5 = 0.98\,\text{m s}^{-2}$. The distance traveled follows from the formula $x = v_0 t + at^2/2 = 0.98 \times 5^2/2 = 12.3\,\text{m}$. Thus the work done is $W = F_x x = 4.92 \times 12.3 = 60.5\,\text{J}$.

S90. The train initially has no kinetic energy ($T_1 = 0$), but eventually acquires a speed of $v = 72\,\text{km/h} = 20\,\text{m s}^{-1}$. It therefore has kinetic energy $T_2 = mv^2/2 = 10^3 \times 10^3 \times 20^2/2 = 2 \times 10^8\,\text{J}$. This energy was all supplied by the motor, which did no other work, so that $W = T_2 - T_1 = 2 \times 10^8\,\text{J}$.

S91. The increase ΔU in the gravitational potential energy is the difference between the energies in the final and initial states. Thus $\Delta U = mgy_2 - mgy_1 = mgh$, where $m = 10$ kg is the mass of the bucket and contents, y_2, y_1 are the final and initial heights of the bucket measured from an arbitrary origin, and $h = 10$ m is their difference. Thus $\Delta U = 10 \times 9.8 \times 10 = 980$ J. The work done against gravity must equal the change of potential energy (there is no kinetic energy in either the initial or final state). Thus $W = \Delta U = 980$ J.

S92. We choose the ground as the zero-point of gravitational potential energy. The total energy of the rollercoaster remains fixed as friction is neglected. Its value can be found at the first point (maximum height) as $E = T_1 + U_1 = mv_1^2/2 + mgh_1$. At the second (minimum height) point the energy is $E = T_2 + U_2 = mv_2^2/2 + mgh_2$. Equating these two expressions we get $v_2^2 = v_1^2 + g(h_1 - h_2)$. Thus

$$v_2 = \sqrt{v_1^2 + 2g(h_1 - h_2)} = \sqrt{0.5^2 + 2 \times 9.8 \times (50 - 5)}$$

i.e. $v_2 = 29.7$ m s^{-1}. Similarly, replacing h_2 by h_3 we get $v_3 = 24.3$ m s^{-1}. Note that the kinetic energy at the highest point is actually negligible.

S93. The law of conservation of energy implies $T_2 + U_2 = T_1 + U_1 - W$, where T_2, U_2 are the ball's kinetic and potential energies at impact, and U_1 is the potential energy of the ball as it is served. Choosing the court level as the zero-point of potential energy, $U_2 = 0, U_1 = mgh$, so $T_2 = T_1 + mgh - W = 10 + 0.06 \times 9.8 \times 2 - 5 = 6.18$ J. Thus using $T_2 = mv_2^2/2$ we get $v_2 = (2T_2/m)^{1/2} = 14.4$ m s^{-1}.

S94. Energy conservation gives $T_2 = T_1 + W$, where T_1, T_2 are the kinetic energies at the beginning and end of the motion, and W is the work done by the resultant external force. The latter is given by $F = ma$ (Newton's second law), so that $W = Fx = max$. Thus $mv_2^2/2 = mv_1^2/2 + max$. With $v_1 = v_0, v_2 = v$ the kinematic formula follows.

S95. Energy conservation implies $T_2 + U_2 = T_1 + U_1$, where subscript 2 refers to the landing platform. With $U_1 = mgh, U_2 = mg \times 1$ and $T_1 = mv_1^2/2$, we get $v_2^2 = v_1^2 + 2g(h - 1) = 3^2 + 2 \times 9.8 \times 1 = 28.6$ m^2 s^{-2}. Thus the impact velocity is $v_2 = \sqrt{28.6} = 5.35$ m s^{-1}.

To find the direction of this velocity, we use the fact that the horizontal velocity is constant. Thus $v_{2x} = v_1$ and since $\cos\theta = v_{2x}/v_2$ (see Figure), this gives $\cos\theta = v_1/v_2 = 3/5.35 = 0.561$, or $\theta = 55.9°$.

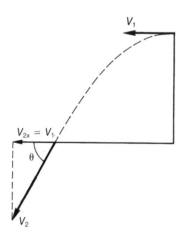

S96. The pole-vaulter's only source of energy is his horizontal kinetic energy T_1. By conservation of energy this must be at least as large as his gravitational potential energy U_2 immediately above the bar. Thus $mv_{\min}^2/2 = U_2 = mgh$. Hence $v_{\min} = \sqrt{2gh} = \sqrt{2 \times 9.8 \times 6.1} = 10.93$ m s^{-1}. This is a minimum since (a) the vaulter has nonzero kinetic energy in traversing the bar, and (b) there are several energy losses, e.g. air resistance, energy losses in deforming the pole (most of the bending work is returned to the vaulter, but not all).

S97. In each pushup the work done by the student's muscles is $W = \Delta U$ where $\Delta U = mgh$. Thus $W = 75 \times 9.8 \times 0.2 = 147\,\text{J}$. Thus, assuming all of this energy is lost, he must perform $1500 \times 1000/147$ or about 10,200 pushups! Do not despair! The largest part of food energy is expended in maintaining the body temperature and body activity such as respiration, etc.

S98. The skier moves a distance $x = vt = 10 \times 10 = 100$ m in 10 s. The work done by the boat's force is thus $W = Fx = 100 \times 100 = 10\,\text{kJ}$. As the velocity v remains constant, the skier's energy does not change. Thus all of this work is lost to frictional forces.

S99. Doubling the speed v quadruples the kinetic energy $mv^2/2$, where m is the car's mass. This energy has to be dissipated by the car's brakes. If the maximum braking torque on the wheels is the same, the wheels will have to revolve four times as far, and hence the car will have to travel four times as far, if it is to stop without skidding.

S100. By energy conservation we have $E_2 = E_1 + W - W_f$. Here W is the work done by the dragging force F, W_f is the work done against friction and $E_2 = mv_2^2/2 + mgy_2$ and $E_1 = mv_1^2/2 + mgy_1$ are the suitcase's final and initial energies. Thus

$$W_f = E_1 - E_2 + W = mg(y_1 - y_2) - \frac{mv_2^2}{2} + W = -mgh - \frac{mv_2^2}{2} + W.$$

Substituting $m = 20$ kg, $v_2 = 1$ m s^{-1}, $h = 5$ m and $W = Fx = Fh/\sin\alpha = 150 \times 5/0.5$ (see Figure in problem), we get

$$W_f = -20 \times 9.8 \times 5 - 20 \times 1^2/2 + 1500 = 510 \text{ J}.$$

Since the frictional force $f = \mu mg \cos\alpha = 20 \times 9.8 \times 0.866\mu = 170\mu$ N, and $W_f = fx = fh/\sin\alpha$, we find $510 = 170\mu 5/0.5 = 1700\mu$. Thus $\mu = 510/1700 = 0.3$.

S101. In the first case (two pulleys) the work done in raising the mass through h is $W = (Mg/2).(2h) = Mgh$, the same as if the mass had been lifted without using the pulleys. Similarly in the second case we get $W = (Mg/4).(4h) = Mgh$.

For the gear wheels, we have a rate of working against the resistive torque which is $P = \Omega.G_2 = (b/a)\omega(a/b)G_1 = \omega G_1$, so the rate of working is the same whichever shaft is turned.

In practice, frictional energy losses can become very severe if there are many moving parts.

S102. Since the motion is uniform, the vertical forces on the load must balance, i.e. $F = mg$, where F is the force exerted by the crane on the load. The power follows from the formula $P = Fv$, giving $P = mgv = 500 \times 9.8 \times 2 = 9800$ W. To find the work done by the crane we use $W = Fh$ since the force is constant. Thus $W = 500 \times 20 = 1000$ J.

Since the second crane lifts the load at twice the above speed, its power is larger by a factor of 2. The work, however, has the same value, since both the force and the height are the same as in the first crane.

S103. The work done by the pump in ejecting a mass m of water is given by conservation of energy: $W = E_2 - E_1 = T_2 - T_1 + U_2 - U_1 = mv_2^2/2 + mgd$, since the water in the well is at rest and we can take its surface as the zero-point of potential energy (we assume that the water level does not change significantly during pumping). Thus the effective power P_{eff} of the pump is given by

$$P_{\text{eff}} = \frac{W}{t} = \frac{m}{t}\left(\frac{v_2^2}{2} + gd\right).$$

The flow rate 2 m³ per second implies $m/t = 2 \times 10^3 \text{ kg s}^{-1}$, since 1 m^3 of water has a mass of 10^3 kg. Substituting also $v_2 = 10 \text{ m s}^{-1}$ and $d = 50$ m, we get $P_{\text{eff}} = 2 \times 10^3(100/2 + 9.8 \times 50) = 1080 \text{ kW}$. The efficiency $\eta = 0.8$ implies that $P_{\text{eff}} = 0.8P$, so that the power consumed is $P = P_{\text{eff}}/0.8 = 1350$ kW.

S104. The car's original kinetic energy is $T = mv^2/2$ has to be dissipated in time t, so the average rate of working of the brakes is $P = mv^2/2t = 10^3 \times (27.8)^2/(2 \times 10) = 38.6$ kW. (100 km/h = 27.8 m s⁻¹.) All of this goes initially into heating the braking surfaces, so they must lose at this rate in order not to heat up.

S105. Animals jumping to the same heights h gain the same potential energy $V/m = mgh/m = gh$ per unit mass. Since their muscle masses scale with their total masses, this suggests that the total energy supplied per unit muscle mass is similar in similar animals. The vertical speed required is similar (of order $(2gh)^{1/2}$), but larger animals need more room to achieve it, suggesting that the *rate* of energy release is lower for larger animals, roughly as l^{-1}, where l is the size.

S106. We write U_g, U_{el} for the gravitational and elastic potential energies. The energy of the mass–spring system is constant. Initially it is $E = U_g = mgh$, since the mass is at rest and the spring is relaxed. On the level surface $E = mv^2/2$, since the mass is at zero height and the spring is still relaxed. Thus $v = \sqrt{2gh}$.

After the mass encounters the spring, it compresses it until all the energy is in the form of elastic potential energy (maximal compression, zero velocity, zero height). Thus $E = U_{el} = kx^2/2$. Equating this to the first expression for E and using $x = h/10$ gives $kh^2/200 = mgh$, i.e. $k = 200mg/h$. Since no energy is lost, the mass returns to exactly the same height h after the spring relaxes.

S107. Energy conservation applied to the motion between the initial (1) and highest (2) positions gives

$$E_1 = E_2 + W,$$

where W is the work done against friction. With $E_1 = mv^2/2$, $E_2 = mgh = mgd \sin \alpha = mgd/\sqrt{2}$, and $W = fd = \mu N d = \mu mgd \cos \alpha = 0.1mgd/\sqrt{2}$ this gives

$$\frac{v^2}{2} = \frac{gd}{\sqrt{2}} + \frac{0.1gd}{\sqrt{2}} = \frac{1.1gd}{\sqrt{2}}.$$

Thus $d = v^2/(1.1\sqrt{2}g)$.

As we saw in P7, the mass can only rest in equilibrium under gravity and friction on an inclined plane if $\mu_s \geq \tan \alpha$. Here $\tan \alpha = 1$, so that the required μ_s is 1. In practice this is impossible.

Using energy conservation for the downward motion gives

$$E_2 = E_3 + W',$$

where E_3 is the energy when the mass returns to its starting point, and W' is the work done against friction on the descent. Because the normal force is the same, the distance traveled is the same, and the coefficient of friction has not changed, $W' = W = 0.1mgd/\sqrt{2} = 0.0454mv^2$, where we have substituted $d = v^2/(1.1\sqrt{2}g)$ from above. Further, $E_2 = mgd/\sqrt{2} = 0.454mv^2$. Thus $E_3 = E_2 - W' = 0.409mv^2$. Equating this to $mv_3^2/2$ gives $v_3 = 0.905v$ for the return velocity. This is smaller than the initial velocity, since energy has been lost performing work against friction.

☐ MOMENTUM AND IMPULSE

S108. Horizontal momentum is conserved as there are no external horizontal forces, i.e. the total momenta before and after the collision are equal. Choosing the bird's motion to define the positive x-direction, we have

$$MV - mv = (M + m)U.$$

Thus $U = (MV - mv)/(M + m)$. For the case given we get $U = (MV - 0.01M \times 10V)/(M + 0.01M) = 0.9MV/1.01M = 0.89V$. Note that energy is *not* conserved in this case: some is lost from the mechanical system.

S109. Conservation of momentum implies $Mv + mu = 0$, where v is the velocity of the gun after firing. Thus $v = -mu/M = -2.3 \, \text{m s}^{-1}$. The minus sign here shows that the gun recoils, with recoil velocity $|v| = 2.3 \, \text{m s}^{-1}$. To reach this speed by being dropped from rest, the kinematic formula $v^2 = v_0^2 + 2ax$ shows that an initial height $h = v^2/2g = 0.28$ m is required.

S110. To achieve the highest terminal velocity, conservation of momentum shows that one needs to maximize the momentum of the exhaust fuel and minimize the final mass of the rocket. Rockets thus use powerful fuels (high exhaust velocity) and carry as large a mass of it as possible. Once a fuel tank is empty, it is jettisoned, reducing the propelled mass and thus raising the final speed.

S111. Momentum conservation gives

$$mu + 0 = mv_1 + mv_2, \tag{1}$$

where v_1, v_2 are the final velocities of the cue ball and pool ball respectively. As there are two unknowns in this equation we must use a second relation. This is supplied by mechanical energy conservation, i.e.

$$m\frac{u^2}{2} + 0 = m\frac{v_1^2}{2} + m\frac{v_2^2}{2}. \tag{2}$$

Rearranging and canceling m we get

$$v_1 - u = -v_2, \tag{3}$$

$$v_1^2 - u^2 = -v_2^2. \tag{4}$$

Dividing (4) by (3), we get

$$v_1 + u = v_2. \tag{5}$$

Adding this to (3) we get $2v_1 = 0$, so $v_1 = 0$. Thus from (5) $v_2 = u$. The cue ball stops dead and the pool ball moves off with the cue ball's original velocity. Note that the restriction to pure sliding motion is unrealistic in practice, as the energy in the rolling motion of the balls is usually significant and causes them to behave differently (see S117).

S112. Momentum conservation gives

$$m_1 u_1 + m_2 u_2 = m_1 v_1 + m_2 v_2.$$

Energy conservation gives

$$\tfrac{1}{2}m_1 u_1^2 + \tfrac{1}{2}m_2 u_2^2 = \tfrac{1}{2}m_1 v_1^2 + \tfrac{1}{2}m_2 v_2^2.$$

We can rewite these equations as

$$m_1(u_1 - v_1) = m_2(v_2 - u_2),$$

$$m_1(u_1 - v_1)(u_1 + v_1) = m_2(v_2 - u_2)(v_2 + u_2).$$

Dividing these equations gives $u_1 + v_1 = v_2 + u_2$, or

$$v_2 - v_1 = -(u_2 - u_1)$$

as required.

S113. We treat both cases simultaneously by writing m for the mass of the incoming particle and u, v for its velocities before and after collision. The proton velocity after collision is v_p. We assume that no external forces act on the particles, and that they collide elastically. Then both momentum and mechanical energy are conserved.

$$mu = mv + m_p v_p \tag{1}$$

and

$$\frac{mu^2}{2} = \frac{mv^2}{2} + \frac{m_p v_p^2}{2} \tag{2}$$

Rearranging we get

$$u - v = \frac{m_p}{m} v_p \tag{3}$$

and

$$u^2 - v^2 = \frac{m_p}{m} v_p^2. \tag{4}$$

Dividing (4) by (3) gives

$$u + v = v_p. \tag{5}$$

Adding (5) and (3) gives $2u = (1 + m_p/m)v_p$. Thus

$$v_p = \frac{2m}{m + m_p} u.$$

Using this in (5) gives

$$v = \frac{m - m_p}{m + m_p} u.$$

Assume $u > 0$ in both cases. In the first collision we have $v = v_1 < 0$. Thus $m = m_1 < m_p$. In the second collision we have $v = v_2 > 0$, so $m = m_2 > m_p$. A lighter particle recoils from a stationary target, while a heavier one moves forward after collision.

Using the last two equations twice, with $m = m_1 = m_p/2$, $m = m_2 = 2m_p$, we get final velocities $v_p = 2u/3$, $v = -u/3$ and $v_p = 4u/3$, $v = u/3$.

S114. From the previous problem, the proton's velocity after collision is

$$v_p = \frac{2m}{m + m_p} u.$$

Its total energy (all kinetic, $= m_p v_p^2/2$) is therefore

$$E_p = \frac{2m_p m^2}{(m + m_p)^2} u^2.$$

All of this energy was transferred from the incoming particle, so $\Delta E = E_p$. The incoming particle had energy $E = mu^2/2$, so

$$\frac{\Delta E}{E} = \frac{4m m_p}{(m + m_p)^2},$$

independent of u. Note that if the incoming particle is an electron, $m = m_e \ll m_p$, so this fraction becomes $\Delta E/E \approx 4m_e/m_p \ll 1$. If the incoming particle is much more massive than the proton, $m \gg m_p$, the transfer is similarly inefficient. Only when the masses are comparable is the transfer significant.

S115. Momentum conservation gives

$$m_1 u_1 = m_1 v_1 + m_2 v_2,$$

and the energy equation is now

$$v_2 - v_1 = eu_1. \tag{1}$$

Eliminating v_1 between these equations gives

$$v_2 = \frac{m_1(1 + e)}{m_1 + m_2} u_1, \tag{2}$$

so that the ratio of the kinetic energy of m_2 after the collision to that of m_1 before it is

$$\frac{\frac{1}{2} m_2 v_2^2}{\frac{1}{2} m_1 u_1^2} = (1 + e)^2 \frac{m_1 m_2}{(m_1 + m_2)^2}.$$

For $m_1 \gg m_2$ this ratio is $(1 + e)^2 m_2/m_1 \ll 1$, and for $m_1 \ll m_2$ it is $(1 + e)^2 m_1/m_2 \ll 1$. (Compare with S114, where $e = 1$.)

S116. We must supply a fixed amount of energy to drive the nail in. From the previous answer we see that energy transfer in a collision is efficient only if the bodies have similar masses. So jumping on a nail wastes a lot of energy. The collision with your shoes is also likely to be more inelastic ($e < 1$) than hammering it, wasting even more energy.

S117. Equation (2) of S115 gives

$$v_2 = \frac{1}{2}(1+e)u_1.$$

Equation (1) of S115 gives

$$v_1 = v_2 - eu_1 = \frac{1}{2}(1-e)u_1$$

for the velocity of the cue ball after the collision, since $m_1 = m_2$. At first sight it appears that the physicist is right, since if e is close to 1, v_1 must be much smaller than v_2. However, the argument is correct only if the cue ball was in pure sliding motion, whereas in reality it is usually rolling. The spin of the ball then causes the ball to continue to move after the collision. (A purely sliding ball stops almost dead at impact – this is a *stun* shot. The ball must be cued at exactly one-half of its height for this to happen. See S211.)

S118. Momentum conservation gives

$$m_1u_1 + m_2u_2 = m_1v_1 + m_2v_2 \tag{1}$$

as before, where v_1, v_2 are the velocities of the bat and ball respectively. We can use the result of the last question to express the elastic (energy conservation) condition as

$$v_2 - v_1 = u_1 - u_2.$$

We wish to find $v_2 = v_1 + u_1 - u_2$, so we need to eliminate the unknown v_1. Using (1) we have

$$v_1 = \frac{m_1u_1 + m_2u_2 - m_2v_2}{m_1},$$

so

$$v_2 = 2u_1 - u_2 + \frac{m_2}{m_1}(u_2 - v_2).$$

For $m_1 \gg m_2$ the term in brackets is negligible, and we get $v_2 = 2u_1 - u_2$. As the term in brackets is negative, this is the maximum value of v_2. Faster pitches can be hit further. However, even the slowest ball is of no use if the hitter's value of $2u_1$ is already large enough for a home run.

S119. Before the man throws the boot, the total momentum of the systems is zero. After the throw it is $mu_1 + Mu_2$, where u_2 is the velocity of the boxcar. Since momentum is conserved we have

$$0 = mu_1 + Mu_2,$$

or $u_2 = -mu_1/M$. The boxcar thus moves (slowly, since $M \gg m$) in the opposite direction from the boot. The boot's velocity *relative* to the boxcar is

$$u_r = u_1 - u_2 = u_1 \left(1 + \frac{m}{M} \right).$$

Hence after a time

$$t = \frac{d}{u_r} = \frac{d}{u_1} \frac{M}{M + m}$$

the boot hits the opposite end of the boxcar. The collision is completely inelastic, so after it the boot and the boxcar move with the same velocity, which we call V. Applying momentum conservation we find

$$(M + m)V = mu_1 + Mu_2 = 0,$$

i.e. the boxcar stops. It has moved a distance

$$l = u_2 t = -\frac{mu_1}{M} \frac{d}{u_1} \frac{M}{M + m} = -\frac{m}{M + m} d$$

in the opposite direction from the boot. As $m \ll M + m$, we have $l \ll d$! this is not a very efficient method of moving the boxcar: the center of mass of the boot–boxcar system has not moved.

S120. The analysis is the same as before until the ball hits the wall. Then we can use the results of S112 to write the momentum conservation equation as

$$mv_1 + Mv_2 = mu_1 + Mu_2 = 0, \tag{1}$$

and energy conservation as

$$v_2 - v_1 = u_1 - u_2. \tag{2}$$

Here u_1, u_2 are the velocity of the ball and boxcar before the collision, and v_1, v_2 the velocities after it. From (1) we have $v_1 = -Mv_2/m$ and $u_1 = -Mu_2/m$, so substituting in (2) we find

$$v_2 + \frac{M}{m} v_2 = -\frac{M}{m} u_2 - u_2,$$

implying $v_2 = -u_2$. The velocities of the boxcar and ball reverse at the collision, so the motion reverses itself. If all subsequent collisions were elastic, the ball and boxcar would oscillate back and forth forever between the original position and some other point. Note that the center of mass of the system remains stationary at all times.

S121. Immediately before hitting the floor for the first time, the ball has velocity u_0 downwards. The coefficient of restitution e is defined so that (velocity of separation) $= e \times$ (velocity of approach). Here the separation velocity is the upward velocity u_1 immediately after the bounce, so that $u_1 = eu_0$. The kinematic formula $v^2 = v_0^2 + 2ax$ now gives the height reached on the first bounce as

$$x_1 = u_1^2/2g = e^2 u_0^2/2g.$$

Clearly, the ball hits the ground for the second time with velocity u_1, and leaves it with upward velocity $u_2 = eu_1$. The same kinematic calculation now shows that the ball reaches a height

$$x_2 = u_2^2/2g = e^4 u_0^2/2g$$

on the second bounce. By the same reasoning, after n bounces the ball reaches height $x_n = e^{2n} x_0$, with $x_0 = u_0^2/2g$.

S122. From the kinematic formula $v = v_0 + at$ with $v = 0, v_0 = u_1$ and $a = -g$ the time to reach the top of the first bounce is $t_1/2 = u_1/g$, so the total time between first and second impacts is $t_1 = 2u_1/g = 2eu_0/g$. After the second impact the upward velocity is $u_2 = eu_1 = e^2 u_0$, so the time between second and third impacts is $t_2 = 2u_2/g = 2e^2 u_0/g$. In an exactly similar way, we see that the time between the nth and $(n+1)$th impact is $t_n = 2e^n u_0/g$. Hence the total time before bouncing stops is

$$t_{\text{bounce}} = \frac{2u_0 e}{g}(1 + e + e^2 + e^3 + \ldots\ldots).$$

The quantity in brackets is an infinite geometric series, whose sum is $(1 - e)^{-1}$. [If this result is unfamiliar, let $S = 1 + e + e^2 + e^3 + \ldots$, then $eS = e + e^2 + e^3 \ldots$, so subtracting we find that $S(1 - e) = 1$, hence the result.] Thus

$$t_{\text{bounce}} = \frac{2u_0}{g}\frac{e}{1 - e}.$$

S123. The highest point is reached when the vertical velocity $v_y = 0$. Using the formula $v_y = v_{y0} - gt$ with $v_{y0} = v_0 \sin\theta$, this happens at time $t_m = v_0 \sin\theta/g$. The corresponding horizontal distance is $x_m = v_{0x} t_m$, since the horizontal motion is uniform. Thus $x_m = v_0^2 \sin\theta \cos\theta/g$ (note that this is half the total range of the shell).

Immediately before the explosion the shell is moving horizontally, and momentum is conserved as it breaks into two. Before the explosion the horizontal momentum of the shell was $p_1 = mv_{0x} = mv_0 \cos\theta$. Since the first shell fragment has zero velocity and thus zero momentum, all of the horizontal momentum must be given to the second fragment. Thus its momentum is $p_2 = p_1 = mv_0 \cos\theta$. As its mass is $m/2$, its horizontal velocity is $u = 2v_0 \cos\theta$.

The vertical velocity of both fragments is still zero after the explosion. Thus they both hit the ground a time t_m after the explosion (i.e. the same as the time to reach the maximum height). After this time the second fragment has moved a horizontal distance $s = ut_m = 2v_0^2 \sin\theta \cos\theta/g$ (note that this is $2x_m$) from the explosion point. It therefore falls at a total distance $X = x_m + s = 3x_m = 3v_0^2 \sin\theta \cos\theta/g$. Thus $X = 3 \times 450^2 \times 0.707^2/9.8 = 30,986$ m.

S124. We can find the horizontal velocity just after impact from the kinematic data. The time required to fall to the ground is $t = \sqrt{2h/g}$, and thus its horizontal velocity must have been $v = x/t = x\sqrt{g/2h}$.

We choose the direction away from the wall as the positive direction. Then the ball's momentum change in the impact is

$$\Delta p = p_2 - p_1 = mv - m(-u) = m(u + v).$$

With the data given, $v = 15 \times \sqrt{9.8/2 \times 4.9} = 15\,\mathrm{m\,s^{-1}}$. Thus $\Delta p = 0.1(20 + 15) = 3.5\,\mathrm{kg\,m\,s^{-1}}$. This is the impulse $I = \Delta p$ provided by the wall.

As the rebound velocity is smaller than the impact velocity, mechanical energy was lost in the impact, which is therefore inelastic.

S125. Momentum is conserved in the impact, so

$$mu = (m + M)V,$$

as the bullet is embedded in the block after impact. Thus the muzzle velocity is

$$u = \left(1 + \frac{M}{m}\right)V = (1 + 7/0.01) \times 0.5 = 350.5 \text{ m s}^{-1}.$$

The energy change is found by comparing kinetic energies after and before impact. Thus $\Delta E = E_2 - E_1 = (m + M)V^2/2 - mu^2/2$. With the data given and u as above, we find $\Delta E = 0.88 - 614.25 = -613.4$ J. The negative sign means that energy is lost. Note that almost all the energy is lost in this case.

S126. Momentum conservation requires that

$$mu = (M + m)V, \tag{1}$$

where V is the velocity of the block and embedded bullet after impact. The latter thus has kinetic energy $T = (M + m)V^2/2$, which is all converted to gravitational potential energy $(M + m)gh$, i.e. $V^2 = 2gh$. Using this in (1) gives

$$u = \frac{(M + m)}{m}(2gh)^{1/2}.$$

With the data given we find $u = 768$ m s^{-1}. The kinetic energy T of the block and bullet can be rewritten, using (1), as

$$T = \frac{m}{M + m}\frac{1}{2}mu^2,$$

so only a fraction $m/M \sim 10^{-3}$ of the bullet's kinetic energy was used to raise the block. Almost all of it ended up heating the block slightly (cf. S114–S116).

S127. Conservation of horizontal momentum gives

$$mu = (m + 8m)V,$$

where V is the velocity of the dart and block after impact (assumed to be almost instantaneous). Therefore $V = u/9$. This is the initial velocity just after impact: the motion of the block and dart is resisted by the spring. Total mechanical energy is conserved in the subsequent compression of the spring, so $E_2 = E_1$, where E_2 is the total energy at maximum compression and E_1 is the kinetic energy just after impact. Thus

$$\frac{1}{2}kx_m^2 = \frac{1}{2}9mV^2,$$

where x_m is the maximum compression of the spring. With V as above, this gives

$$x_m = \sqrt{\frac{m}{k}}\frac{u}{3}.$$

S128. The locomotive must expend more power because the accumulating snow increases the mass and hence the momentum of the train. In a very short interval Δt, the accumulated mass is $\Delta m = r_m \Delta t$. Hence the momentum change of the train is $\Delta p = \Delta(mv) = r_m \Delta t\, v$. The extra force the locomotive must exert is thus

$$F = \frac{\Delta p}{\Delta t} = r_m v.$$

The power required to maintain the constant speed v with this F is $P = Fv = r_m v^2$. With $v = 108$ km/h $= 30$ m s^{-1}, and $r_m = 10$ kg s^{-1}, we find $P = 9000$ W $= 9$ kW.

S129. We choose the downward vertical as the positive direction. The velocity of the sack just before impact is given by the free-fall formula $v = \sqrt{2gh}$. Since the sack comes to a stop, its entire momentum is lost. Thus the momentum change of the sack is

$$\Delta p = p_2 - p_1 = 0 - Mv = -Mv.$$

Hence the impulse on the sack is $J_s = \Delta p = -Mv$. The impulse on the platform is $J_p = -J_s = Mv$ (Newton's third law). From the data given $J_p = M\sqrt{2gh} = 10 \times \sqrt{2 \times 9.8 \times 1} = 44.3$ kg m s^{-1}.

The average force on the platform follows from $F\Delta t = J$, where Δt is the duration of the impact. With $\Delta t = 0.1$ s and $J = J_p$ as above, we find the average force on the platform $F_p = J_p/\Delta t = 44.3$ N.

Momentum conservation is *not* violated here: the sack and the Earth share the final momentum. Because the mass of the Earth is so high, the recoil is negligible. Momentum is *always* conserved in collision problems; mechanical energy need *not* be (as here).

S130. As in the previous problem, the impact velocity is $v = \sqrt{2gh}$. Thus the momentum of each grain changes by $-mv$ on landing; the momentum of the platform therefore changes by mv as each grain lands. Denote by R the number rate at which grains are deposited on the platform. The corresponding rate of momentum deposition is $\Delta p/\Delta t = Rmv$, and this is therefore the impact force F exerted by the stream of grain. With the data given $F = Rm\sqrt{2gh} = (1000 \times 0.01)\sqrt{2 \times 9.8} = 44.3$ N.

S131. We take the positive direction as that away from the goalkeeper. The momentum change of the ball during impact is $\Delta p = p_2 - p_1 = m_b v - m_b(-u) = m_b(u + v)$. This is the impulse J_b on the ball. The impulse on the goalkeeper is equal and opposite, i.e. $J_g = -J_b = -m_b(u + v)$. Thus the force exerted on the goalkeeper during the punch is $F_g = J_g/\Delta t = -m_b(u + v)/\Delta t$.

If the goalkeeper is not to slide backwards, the resultant force on him immediately after the punch must be zero. Thus $f + F_g = 0$, where f is the frictional force. Thus $f = -F_g = m_b(u + v)/\Delta t$. Since $f < \mu_s m_g g$, we require $\mu_s m_g g > m_b(u + v)/\Delta t$. Rearranging, this gives

$$\mu_s > \frac{m_b}{m_g}\frac{(u + v)}{g\Delta t} = 1.8\frac{m_b u}{m_g g\Delta t},$$

because $v = 0.8u$. With the data given this implies

$$\mu_s > 1.8 \times 0.5 \times 1/(80 \times 9.8 \times 0.2) = 0.006.$$

This small value is reasonable, as in practice goalkeepers do not slide backwards after punching the ball.

S132. We take the boat to move off in the positive direction. The momentum of the boat–man–load system is conserved. Before the first throw it was zero. Thus

$$0 = (M - m)U_1 + m(-v_r).$$

Here $M = M_b + 10m$ is the total mass of the boat–man–load system and U_1 is the boat's velocity after the first throw. Hence

$$U_1 = \frac{m}{M - m} v_r. \tag{1}$$

The second throw takes place from a boat moving with velocity U_1, and mass $M - m$. Thus conservation of momentum implies

$$(M - m)U_1 = (M - 2m)U_2 + m(-v_r + U_1). \tag{2}$$

Here U_2 is the velocity of the boat after the second throw, when its mass is $M - 2m$, and we must use the velocity of the sack relative to the Earth, i.e. $-v_r + U_1$, in writing down its momentum. From (2) we get

$$U_2 = \frac{(M - m)U_1 - m(U_1 - v_r)}{M - 2m}. \tag{3}$$

Substituting U_1 from (1) gives

$$U_2 = \frac{(2M - 3m)m}{(M - m)(M - 2m)} v_r.$$

With the data given $M = M_b + 10m = 200 + 50 = 250$ kg, so

$$U_1 = \frac{5}{245} v_r = 2.04 \times 10^{-2} v_r$$

$$U_2 = \frac{(500 - 15)5}{(250 - 5)(250 - 10)} v_r = 4.12 \times 10^{-2} v_r.$$

Note that U_2 exceeds $2U_1$: the second throw is from a lighter boat and thus gives a higher velocity.

S133. Since momentum is a vector quantity, in a two-dimensional problem both components are conserved. We choose the x-axis in the direction of motion of the first car, and the y-axis in the direction of motion of the second car. Let θ be the angle made by motion of the combined wreck with respect to the x-axis, and V its velocity (see Figure).

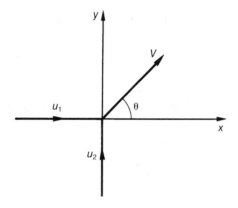

Conservation of momentum in the x-direction gives

$$m_1 u_1 = (m_1 + m_2) V \cos \theta. \tag{1}$$

Conservation of momentum in the y-direction gives

$$m_2 u_2 = (m_1 + m_2) V \sin \theta. \tag{2}$$

We get a relation giving θ by dividing (2) by (1):

$$\tan \theta = \frac{m_2 u_2}{m_1 u_1}.$$

(Note that this is the ratio of the initial momenta.) With the data given, the rhs $= 1$, so $\theta = 45°$. From (1)

$$V = \frac{m_1}{(m_1 + m_2) \cos \theta} u_1$$

i.e. $V = 0.943 u_1 = 17.0$ km/h.

The total initial mechanical energy consists of the kinetic energy of the two cars, and is

$$E_1 = \tfrac{1}{2} m_1 u_1^2 + \tfrac{1}{2} m_2 u_2^2,$$

while the final energy is the kinetic energy of the wreck, i.e.

$$E_2 = \frac{1}{2} (m_1 + m_2) V^2.$$

Thus the mechanical energy change in the collision is

$$\Delta E = E_2 - E_1 = \tfrac{1}{2} m_1 (V^2 - u_1^2) + \tfrac{1}{2} m_2 (V^2 - u_2^2).$$

We need the velocities in m s^{-1}: $u_1 = 18$ km/h $= 5$ m s^{-1}; $u_2 = 36$ km/h $= 10$ m s^{-1}; and $V = 17.0$ km/h $= 4.72$ m s^{-1} The masses are $m_1 = 1000$ kg, $m_2 = 500$ kg. Thus the energy change is

$$\Delta E = 500(4.72^2 - 5^2) + 250(4.72^2 - 10^2) = -20{,}791 \text{ J} \approx -20.8 \text{ kJ}.$$

Most of this energy goes into deforming the cars.

S134. We take the x-direction in the direction of the cue ball's original motion, and the y-direction at right angles to it (see Figure). Let the cue ball's approach velocity be u and the velocities of the cue ball and object ball after collision be v_1, v_2.

Conservation of x-momentum gives

$$mu = mv_1 \cos\theta + mv_2 \cos\phi \tag{1}$$

and conservation of y-momentum gives

$$0 = mv_1 \sin\theta - mv_2 \sin\phi. \tag{2}$$

Conservation of energy (all kinetic) gives

$$\frac{1}{2}mu^2 = \frac{1}{2}mv_1^2 + \frac{1}{2}mv_2^2. \tag{3}$$

Note that the mass m of each ball cancels from all of the equations. From (2) we get

$$v_1 \sin\theta = v_2 \sin\phi, \tag{4}$$

so eliminating v_2 from (1) gives

$$u = v_1 \cos\theta + \frac{v_1 \sin\theta \cos\phi}{\sin\phi}.$$

Thus

$$u = \frac{v_1 \sin(\theta + \phi)}{\sin\phi}, \tag{5}$$

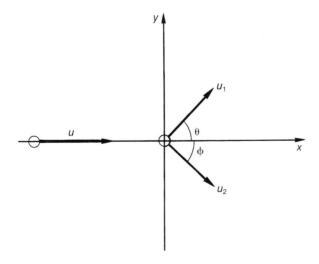

where we have used the trigonometric identity $\sin(\theta + \phi) = \sin\theta\cos\phi + \cos\theta\sin\phi$. Now substituting for v_2 and u from (4) and (5) in (3) gives

$$v_1^2 + v_1^2 \frac{\sin^2\theta}{\sin^2\phi} = v_1^2 \frac{\sin^2(\theta + \phi)}{\sin^2\phi},$$

or

$$\sin^2\theta + \sin^2\phi = \sin^2(\theta + \phi).$$

This equation is satisfied if

$$\theta + \phi = 90°,$$

since then the lhs is $\sin^2\theta + \cos^2\theta = 1$, and the rhs is $\sin 90° = 1$.

S135. We choose the x-axis in the direction of motion of the first fragment. Denoting the masses as $m_1, m_2, m_3 = M - m_1 - m_2$ (see Figure) we can write the equations of momentum conservation as

$$0 = m_1 v - m_2(2v)\sin 30° - (M - m_1 - m_2)(3v)\sin 30°, \qquad (1)$$

$$0 = m_2(2v)\cos 30° - (M - m_1 - m_2)(3v)\cos 30°. \qquad (2)$$

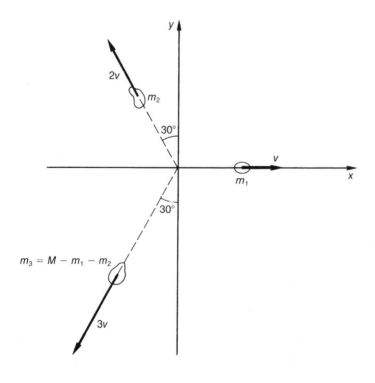

Using $\sin 30° = 0.5$, and canceling v, we find from (1) that

$$m_1 - m_2 - 1.5(M - m_1 - m_2) = 0,$$

or

$$5m_1 + m_2 = 3M. \tag{3}$$

Canceling $v\cos 30°$ from (2), we get

$$3(M - m_1 - m_2) - 2m_2 = 0,$$

or

$$3m_1 + 5m_2 = 3M. \tag{4}$$

The solution of the simultaneous equations (3, 4) is $m_1 = 6M/11$, $m_2 = 3M/11$, so $m_3 = 2M/11$.

The total mechanical energy after the explosion is

$$E = \tfrac{1}{2}(m_1 v^2 + m_2(2v)^2 + m_3(3v)^2).$$

Substituting the above values we get

$$E = \frac{6 + 3 \times 4 + 2 \times 9}{22} Mv^2 = \frac{18}{11} Mv^2.$$

☐ CIRCULAR AND HARMONIC MOTION

S136. The radius of the spaceship's orbit is $r = R + h = 5200$ km $= 5.2 \times 10^6$ m. The tangential velocity is given by the circumference of the circle divided by the time $P = 2$ h$= 7200$ s $= 7.2 \times 10^3$ s to travel around it, i.e.

$$v = \frac{2\pi r}{P} = 4.54 \times 10^3 \, \text{m s}^{-1}.$$

The angular velocity follows from the definition

$$\omega = \frac{v}{r} = 8.73 \times 10^{-4} \, \text{rad s}^{-1}.$$

The centripetal force is given by the formula

$$F = m\frac{v^2}{r}.$$

Substituting $m = 10^4$ kg and the values of v and r, we get $F = 3.96 \times 10^4$ N.

S137. The tension T in the string is the centripetal force required to keep the car in circular motion. Using the formula

$$F = m\frac{v^2}{r} = m\omega^2 r$$

and substituting the values of m, ω and r we get $T = F = 0.1$ N.

S138. The forces acting on the plumbline bob (mass m) are its weight mg and the tension T of the string. The resultant of these must provide the centripetal force $F_c = mR\omega^2 \cos\lambda$ towards the Earth's axis (see Figure). Taking the x- and y-directions along the local horizontal (North) and vertical (towards the center of the Earth) we have in the x-direction

$$\Sigma F_x = T\sin\theta,$$

and in the y-direction

$$\Sigma F_y = mg - T\cos\theta.$$

In the x, y system, the centripetal force has components

$$F_{cx} = mR\omega^2 \cos\lambda \sin\lambda,$$

$$F_{cy} = mR\omega^2 \cos^2\lambda.$$

Hence setting $\Sigma F_x = F_{cx}, \Sigma F_y = F_{cy}$ gives

$$T\sin\theta = mR\omega^2 \cos\lambda \sin\lambda,$$

$$mg - T\cos\theta = mR\omega^2 \cos^2\lambda.$$

Eliminating T between these two equations gives

$$\tan\theta = \frac{R\omega^2 \cos\lambda \sin\lambda}{g - R\omega^2 \cos^2\lambda}.$$

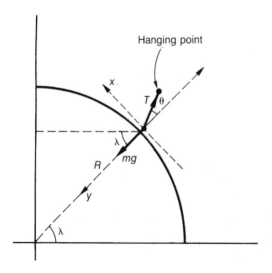

Since $g \gg R\omega^2$ we can drop the second term in the denominator, and we see that $\tan \theta$ is small. Thus $\tan \theta \approx \theta$ if θ is expressed in radians, and we find

$$\theta \approx \frac{R\omega^2 \cos \lambda \sin \lambda}{g} = \frac{R\omega^2}{2g} \sin 2\lambda.$$

Clearly this deflection is always towards the equator, and is a maximum at $\lambda = 45°$. The maximum value is $\theta_{max} = R\omega^2/2g = 1.7 \times 10^{-3}$ rad $= 0.1°$. (The Earth's rotation rate is $\omega = 2\pi/(24 \times 3600) = 7.3 \times 10^{-5}$ rad s^{-1}.)

S139. The centripetal force required to keep the car in circular motion is $F = mv^2/r$. This force must be supplied by static friction. Thus $f = F = mv^2/r$. As the static friction force has a maximum value of $\mu_s mg$, the driver needs

$$\mu_s mg > m\frac{v^2}{r}$$

to stay on the road. Thus μ_s must be larger than $v^2/(rg)$. With $v = 80$ km/h $= 22.2$ m s^{-1} and $r = 100$ m, we find $v^2/(rg) = 0.5$. As this exceeds μ_s the centripetal force is insufficient and the car will leave the road because of its inertia.

S140. Let the mass be m and its speed be v. The resultant force on the mass in its motion must be just the centripetal force required to keep it in its circular trajectory, i.e.

$$\Sigma F = m\frac{v^2}{r},$$

acting in the radial direction (see Figure). At the lowest point

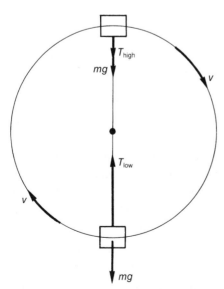

$$\Sigma F = T_{\text{low}} - mg, \tag{1}$$

while at the highest point

$$\Sigma F = T_{\text{high}} + mg. \tag{2}$$

The value of ΣF is the same in both equations, so subtracting (2) from (1) gives $T_{\text{low}} - T_{\text{high}} = 2mg$.

This is independent of the speed and the radius of the circle.

S141. When the string makes an angle θ to the vertical (see Figure) the centripetal force is $T + mg\cos\theta$. This must be constant in uniform circular motion, so the minimum tension T is reached when $\cos\theta$ has its maximum value 1, i.e. at the highest point. Here

$$T + mg = m\frac{v^2}{r}.$$

To keep the string taut requires $T > 0$, i.e. $v^2 > rg$, or

$$v > \sqrt{rg}.$$

With the data given the velocity must exceed 3.13 m s^{-1}.

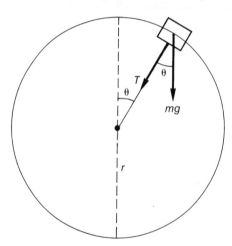

S142. When the string breaks the mass is moving horizontally, so by Newton's first law it will initially continue to do so, with the velocity v it had before the string broke. Thereafter it will fall under gravity and hit the ground. In a recent survey, U.S. college students were asked a similar question. A *majority* (including many science majors) believed that the mass would initially fly radially outwards along the line of the string (here vertically downwards)! Surveys in other countries give similar results. Remember, the string tension is *not* resisting a tendency of the mass to fly radially outwards, but forcing the mass to deviate from its tendency to move in a straight line (tangent to the circle).

S143. Let the velocity of the mass be v at angle θ. Then it has kinetic energy $Mv^2/2$; this must be less than at the lowest point, because some kinetic energy has been converted to potential energy. Conservation of energy gives explicitly

$$\frac{1}{2}Mv_0^2 = \frac{1}{2}Mv^2 + Mg(L - L\cos\theta) \tag{1}$$

(see Figure). The net inward force on the mass must supply the centripetal force, so

$$T - Mg\cos\theta = \frac{Mv^2}{L}.$$

Substituting for v^2 from (1) gives

$$T = M\left(3g\cos\theta - 2g + \frac{v_0^2}{L}\right)$$

as required.

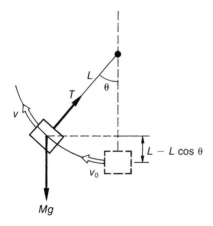

S144. The vertical component of the tension balances the bob's weight and the horizontal component supplies the centripetal force to keep the bob rotating in a horizontal circle of radius R (see Figure). Thus

$$T\cos\alpha = mg \tag{1}$$

and

$$T\sin\alpha = m\frac{v^2}{R}. \tag{2}$$

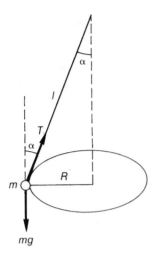

But the bob's speed is $v = 2\pi R f$, and $R = l\sin\alpha$, so (2) gives

$$T\sin\alpha = 4\pi^2 f^2 mR = 4\pi^2 f^2 ml\sin\alpha,$$

so that $T = 4\pi^2 f^2 ml = 158$ N.

Equation (1) now gives $\cos\alpha = mg/T = 0.031$ so that $\alpha = 88.2°$, i.e. the pendulum is almost horizontal.

S145. Clearly the cars are in most danger of falling from the circular loop at its highest point (see Figure). There

$$N + mg = m\frac{v^2}{R}, \tag{1}$$

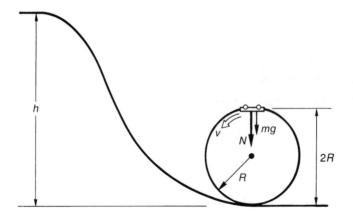

where v is the velocity at this point and N is the track's force on the car; this is normal to the track as there is no friction. In (1) v must be large enough to make N positive, or the cars will detach from the track. Thus we require

$$v^2 > Rg. \tag{2}$$

Mechanical energy is conserved, so equating its values at the high point h and the top of the loop, we get

$$mgh = \frac{1}{2}mv^2 + 2mgR,$$

or

$$h = \frac{v^2}{2g} + 2R.$$

By (2), $h > R/2 + 2R = 2.5R$.

In practice h must be appreciably higher, because of frictional losses.

S146. The forces on the bobsleigh are shown in the Figure. The resultant vertical force must vanish, so that

$$\Sigma F_y = N\cos\alpha - mg = 0, \tag{1}$$

where N is the force exerted by the track on the bobsleigh (normal to its surface as there is no friction) and m is the mass of the bobsleigh. The resultant horizontal force must supply the centripetal force required to keep the bobsleigh in circular motion. Thus

$$\Sigma F_x = N\sin\alpha = m\frac{v^2}{r}. \tag{2}$$

Eliminating N we get

$$v^2 = rg\tan\alpha.$$

With the data given, the maximum $v = 13.0$ m s^{-1}.

If the speed exceeds this value, the bobsleigh moves outwards and therefore

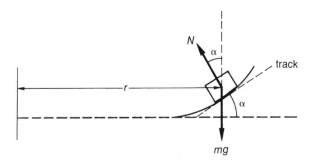

upwards. If the speed is less than this value, it descends to the flat part of the track.

S147. In a turn of radius r the plane and pilot experience centripetal acceleration $a = v^2/r$. If this is to be $\leq 6g$, the smallest r can be is

$$r_{\min} = \frac{v^2}{6g} = \frac{c_s^2}{6g} M^2 = 1.97 M^2 \text{ km}.$$

Thus for $M = 2$, $r_{\min} = 7.9$ km, and for $M = 3$, $r_{\min} = 17.7$ km.

S148. The lift force L must balance the airplane's weight in the vertical direction (no resultant force) and supply the centripetal acceleration in the horizontal direction. If the angle of banking is θ this requires (see Figure)

$$L \cos \theta = Mg,$$

$$L \sin \theta = Ma,$$

where a is the centripetal acceleration v^2/r. Dividing, we get $\tan \theta = a/g$. Since $a = 6g$ in the tightest turns, the banking angle is $\theta = \tan^{-1} 6 = 80.5°$. The pilot's apparent weight is the force N his seat exerts on him. Vertical force balance requires $N \cos \theta = mg$ where m is his mass, so that $N = mg/\cos \theta$. His apparent weight is thus $mg/\cos \theta = 394 \times 9.8 = 3870$ N.

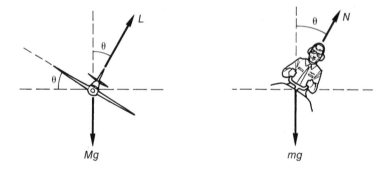

S149. We require the centripetal acceleration $a = v^2/r$ to be less than $0.05g$, so the maximum allowable speed is $v_{\max} = (0.05gr)^{1/2} = 160$ km/h. With the same limit on a and $v = 400$ km/h, we require radii of curvature no less than $r_{\min} = v^2/0.05g = 25$ km. The track must be extremely straight. This creates difficulties in planning routes through populated areas.

S150. The water rises in the outer side of the glasses because excess pressure is needed on that side to supply the centripetal acceleration (see Figure). Consider the point A at the outer side of the glass, level with the water on the

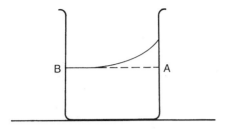

inner side (point B). At B the water pressure is equal to the atmospheric value P_0. To supply the centripetal acceleration to a horizontal column of water of unit cross-sectional area requires a pressure

$$P = P_0 + \rho d a$$

where ρ is the water density (pressure = force per unit area). To maintain the vertical balance of the water above A (height h) requires

$$P = P_0 + \rho g h.$$

Eliminating $P - P_0$ between these equations shows that $da = gh$. With $a \leq 0.05g$ we find $h \leq 0.05d = 0.4\,\text{cm}$. Of course it would be advisable to allow more room between water and brim than this to cover other possible disturbances.

S151. The resultant horizontal force on the mass on the turntable must equal the centripetal force $m\omega^2 r$. At r_{max} the frictional force f opposes the tendency to move outwards (see Figure), so

$$T + f = m\omega^2 r_{max}, \qquad (1)$$

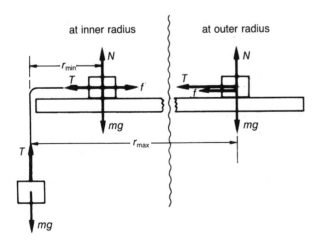

where T is the tension in the string. The latter must equal the weight of the hanging mass, i.e. $T = mg$, while $f = \mu_s mg$. Substituting in (1) we get

$$mg + \mu_s mg = m\omega^2 r_{max},$$

so that

$$r_{max} = \frac{g}{\omega^2}(1 + \mu_s).$$

At $r = r_{min}$ the mass on the turntable is on the verge of moving inwards (see Figure), so that the frictional force is reversed as compared to (1), i.e.

$$T - f = m\omega^2 r_{min}.$$

Substituting $T = mg, f = \mu_s mg$ as before we find

$$r_{min} = \frac{g}{\omega^2}(1 - \mu_s).$$

With the data given we find $r_{max} = (9.8/36)(1 + 0.5) = 0.41 \, \text{m}$, $r_{min} = (9.8/36)(1 - 0.5) = 0.14 \, \text{m}$.

S152. In the case of no friction, the resultant horizontal and vertical forces on the cycle and rider are (see Figure)

$$\Sigma F_x = N \sin \alpha,$$

$$\Sigma F_y = N \cos \alpha - mg,$$

where N is the normal force exerted by the track on the cycle tires. To supply the centripetal force as the cycle performs the turn requires $\Sigma F_x = mv_0^2/r$, while ΣF_y must vanish as there is no vertical motion. Thus

$$N \sin \alpha = m\frac{v_0^2}{r}, \tag{1}$$

$$N \cos \alpha = mg. \tag{2}$$

Dividing (1) by (2) gives $\tan \alpha = v_0^2/(rg)$, so that $v_0 = (rg \tan \alpha)^{1/2}$.

At speed $v_1 = 2v_0$ the cycle and rider are in danger of sliding upwards, so the frictional force f acts downwards (see Figure). Thus

$$\Sigma F_x = N \sin \alpha + f \cos \alpha = m\frac{v_1^2}{r}, \tag{3}$$

$$\Sigma F_y = N \cos \alpha - f \sin \alpha - mg = 0. \tag{4}$$

From (4) we have $N = f \tan \alpha + mg/(\cos \alpha)$, so substituting into (3) we find

$$f \tan \alpha \sin \alpha + mg \tan \alpha + f \cos \alpha = m\frac{v_1^2}{r},$$

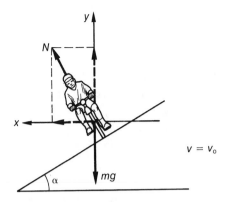

$v = v_o$

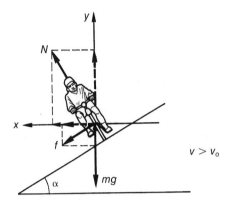

$v > v_o$

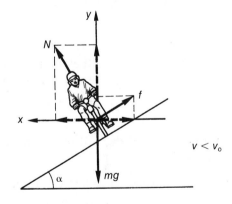

$v < v_o$

i.e.

$$f\left(\frac{\sin^2 \alpha}{\cos \alpha} + \cos \alpha\right) = m\frac{v_1^2}{r} - mg\tan\alpha.$$

Using the trigonometric identity $\sin^2 \alpha + \cos^2 \alpha = 1$, the coefficient of f in this equation is $1/(\cos\alpha)$, so we get

$$f = m\frac{v_1^2}{r}\cos\alpha - mg\sin\alpha.$$

Substituting $v_1 = 2v_0 = 2(rg \tan \alpha)^{1/2}$ we get

$$f = 4mg \sin \alpha - mg \sin \alpha = 3mg \sin \alpha.$$

When the speed is $v_2 = v_0/2$, the cycle and rider are in danger of sliding down the banking so the frictional force f acts upwards (see Figure). Thus

$$\Sigma F_x = N \sin \alpha - f \cos \alpha = m\frac{v_2^2}{r},$$

$$\Sigma F_y = N \cos \alpha + f \sin \alpha - mg = 0.$$

Eliminating N between these two equations similarly as in the previous case, we get

$$f = mg \sin \alpha - m\frac{v_2^2}{r}\cos \alpha;$$

substituting $v_2 = v_0/2 = (rg \tan \alpha)^{1/2}/2$ now gives

$$f = mg \sin \alpha - \tfrac{1}{4}mg \sin \alpha = \tfrac{3}{4}mg \sin \alpha.$$

Note that to find the *coefficient* of friction we would have to divide the expressions for f by those for N in each case.

S153. If the satellite has mass m and speed v its weight mg must supply the centripetal acceleration mv^2/R_e, so that $v = (gR_e)^{1/2}$. The period is $2\pi R_e/v = 2\pi (R_e/g)^{1/2} = 85$ min. Typically the period for low-Earth-orbit satellites is nearer to 90 min.

S154. No! The maximum controlled deceleration a of the car is given by the kinematic formula $v^2 = v_0^2 + 2ax$ as $a = -v_0^2/2r$. To turn the car in a curve of radius r requires centripetal acceleration $-v_0^2/r$, i.e twice as much. (Clearly turning the car also introduces additional risks such as skidding and overturning.)

S155. The period of the pendulum is $P = 2\pi(l/g)^{1/2}$. With $l = 1$ m we find $P = 2$ s, so it performs 1800 swings in one hour.

S156. Accelerating the elevator upwards by a increases the effective gravity g_{eff} to $g + a$ (see S48 or S72). The pendulum period is proportional to $g_{\text{eff}}^{-1/2}$ and therefore shortens. The reverse happens if the elevator accelerates downwards.

S157. Hooke's law states that the force F exerted when the spring extension is x is $F = -kx$. Here this becomes $mg = k\Delta x$, so the spring constant $k = mg/\Delta x = 98 \, \text{N m}^{-1}$. The period of the system is $P = 2\pi(m/k)^{1/2} = 0.63$ s.

S158. The students should first measure the spring constant by hanging a mass m from it. As in the previous answer they get $k = mg/\Delta x$, and the mass–spring system has period $P = 2\pi(m/k)^{1/2} = 2\pi(\Delta x/g)^{1/2}$. A pendulum formed by hanging a mass from the string has period $P' = 2\pi(l/g)^{1/2}$. They must arrange the string length exactly equal to the spring extension, if possible.

S159. The motion is described by

$$x(t) = x_0 \cos \omega t,$$

where $\omega = (k/m)^{1/2}$. (Note that we must express ωt in radians here.) It thus reaches x_1 at time $t_1 = \omega^{-1} \cos^{-1}(x_1/x_0)$. With the data given we find $t_1 = 0.87$ s.

The velocity follows from energy conservation:

$$\tfrac{1}{2}kx_0^2 = \tfrac{1}{2}kx_1^2 + \tfrac{1}{2}mv^2,$$

so that $v = [k(x_0^2 - x_1^2)/m]^{1/2} = 0.16$ m s^{-1}.

S160. Energy conservation can be expressed as

$$v^2 + \omega^2 d^2 = C,$$

where ω is the angular frequency, d is the distance traveled by the end of the pendulum and C is a constant. Since $v = v_0$ when $d = 0$, and $v = 0$ when $d = A$ (the amplitude) we have $A = v_0/\omega = v_0\sqrt{l/g} = 0.2$ m.

S161. Energy conservation requires that $E = mv^2/2 + kx^2/2$ remain constant. Thus

$$mv_1^2 + kx_1^2 = mv_2^2 + kx_2^2,$$

so that $m = k(x_2^2 - x_1^2)/(v_1^2 - v_2^2) = 0.02$ kg. The amplitude is given by setting $v_2 = 0, x_2 = A$, so that $kA^2 = mv_1^2 + kx_1^2$, leading to $A = 0.22$ m with the data given.

S162. The four springs can act together as a single spring of constant $4k$ and thus oscillate at frequency

$$\nu = \frac{1}{2\pi}\left(\frac{4k}{M}\right)^{1/2}.$$

We must ensure that this is smaller than $\nu_m = 10\,\text{s}^{-1}$, so we require $k < 100\pi^2 M = 4935\,\text{N}\,\text{m}^{-1}$. Other modes of oscillation (e.g. rocking) will have lower frequencies, so this is the required limit.

S163. The two springs behave as one spring of constant $k = k_1 + k_2 = 3$ N m^{-1}. The maximum compression of spring 1 occurs after 3/4 of an oscillation period, i.e. after a time $3P/4 = (3\pi/2)(M/k)^{1/2} = 2.7$ s. The maximum compression is the amplitude A, which from energy conservation (see S160) is $A = v_1/\omega = v_1(M/k)^{1/2} = 0.29$ m.

S164. The motion of the mass is given by

$$x(t) = A \sin \omega t$$

with ωt in radians. Here $A = 0.29$ m (see previous answer), and $\omega = (k/M)^{1/2} = 1.73\,\text{rad s}^{-1}$. Hence the time at which $x = -0.1$ m is $t = (1/\omega)\sin^{-1}(x/A)$. Because $x < 0$ we have to convert the negative value of $\theta = \sin^{-1}(x/A)$ (in radians) to $2\pi - |\theta|$. With the data given we find $t = 3.42\,\text{s}$.

S165. Two springs connected "in series" in this way have an effective constant k' given by

$$\frac{1}{k'} = \frac{1}{k} + \frac{1}{k} = \frac{2}{k},$$

so $k' = k/2$. The oscillation period $P = 2\pi(m/k)^{1/2}$ changes to $P' = 2\pi(m/k')^{1/2} = \sqrt{2}P$.

S166. The oscillation frequency is $\omega = 2\pi\nu = (k/m_{\text{tot}})^{1/2}$, where $m_{\text{tot}} = m + M = 5m$ is the total oscillating mass. Thus here $k = \omega^2 m_{\text{tot}} = 4\pi^2\nu^2 \times 5m$, which gives $k = 197\ \text{N m}^{-1}$.

The maximum horizontal force is exerted on the block when the acceleration a is a maximum, which happens at $x = \pm A$. Then $|a|_{\text{max}} = kA/5m$, and we have $F_{\text{max}} = M|a|_{\text{max}} = 4kA/5$. For the case $A = 0.1$ m given this implies $F_{\text{max}} = 15.8$ N.

In all cases this force must be supplied by friction, f, i.e. $f = 4kA/5$. But f is limited by $f \le \mu_s Mg = 4\mu_s mg$, so the maximum possible amplitude A_m is given by

$$\frac{4kA_m}{5} = 4\mu_s mg$$

or $A_m = 5m\mu_s g/k = 0.174$ m.

☐ GRAVITATION

S167. From the formula

$$F = \frac{GM_1 M_2}{d^2}$$

with $M_1 =$ Sun's mass, $M_2 =$ Earth's mass, and the data given, we find $F = 6.7 \times 10^{-11} \times 2 \times 10^{30} \times 6 \times 10^{24}/(1.5 \times 10^{11})^2 = 3.57 \times 10^{22}$ N.

S168. The planet's angular velocity is $\omega = 2\pi/P$. If the planet has mass m, the gravitational force $F = GM_\odot m/a^2$ must supply the centripetal force $F_c = ma\omega^2 = ma(2\pi/P)^2$ required to keep it in a circular orbit. Equating F, F_c gives

$$a^3 = \frac{GM_\odot P^2}{4\pi^2}.$$

Thus the planet's year is

$$P = \frac{2\pi}{(GM_\odot)^{1/2}} a^{3/2}.$$

This relation is true even if the planet's orbit is elliptical and a is the semi-major axis (in practice all planetary orbits are slightly elliptical), and is known as Kepler's third law.

S169. By definition the weight is equal to the normal force N that must be exerted by the Earth's surface on the mass in equilibrium. At the equator the resultant force on a mass m is

$$\Sigma F_x = F_g - N,$$

where F_g is the gravitational force on the mass (see Figure). This must supply the centripetal force $m\omega^2 R_e$ needed to keep the mass in circular motion with angular velocity ω. Substituting $F_g = GM_e m / r_e^2$, we find

$$N = F_g - \Sigma F_x = \frac{GM_e m}{R_e^2} - m\omega^2 R_e,$$

where M_e is the Earth's mass. By definition $g_{\text{eff}} = N/m$, so at the equator

$$g_{\text{eff}}(\text{eq}) = \frac{GM_e}{R_e^2} - \omega^2 R_e. \tag{1}$$

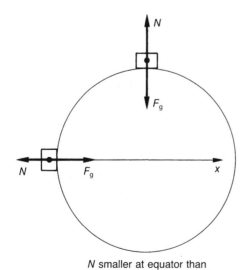

N smaller at equator than
at pole

At the pole no centripetal acceleration is required, so

$$\Sigma F_x = F_g - N = 0.$$

Thus

$$g_{\text{eff}}(\text{pole}) = \frac{N}{m} = \frac{GM_e}{R_e^2}. \tag{2}$$

Dividing (1) by (2) gives

$$\frac{g_{\text{eff}}(\text{eq})}{g_{\text{eff}}(\text{pole})} = 1 - \frac{\omega^2 R_e^3}{GM_e}.$$

With $\omega = 2\pi/24 \, \text{rad h}^{-1} = 7.27 \times 10^{-5} \, \text{rad sec}^{-1}$ and the data given we find

$$\frac{g_{\text{eff}}(\text{eq})}{g_{\text{eff}}(\text{pole})} = 1 - 3.44 \times 10^{-3} = 0.997.$$

The effective gravity hardly differs at the poles and the equator because the Earth rotates comparatively slowly.

S170. If g_{eff} vanishes at the equator, we have

$$\frac{GM}{R^2} = \omega^2 R$$

[cf. equation (1) of the previous solution], where $\omega = 2\pi/P_b$. Thus

$$P_b = 2\pi \left(\frac{R^3}{GM} \right)^{1/2}. \tag{1}$$

For the Earth we find $P_b = 1.4$ h. (Note that we can also write $P_b = (3\pi/G\rho)^{1/2}$, where $\rho = 3M/4\pi R^3$ is the average density of the body.)

S171. If the period is shorter than P_b, gravity will be unable to supply the required centripetal acceleration at the equator, and matter from the star will fly off. The star will be disrupted dynamically at any period significantly shorter than P_b.

Using the expression (1) of the previous solution for P_b, we find $P_b = 6$ s, $P_b = 5 \times 10^{-4}$ s for the white dwarf and neutron star respectively. Since we must have $P_p > P_b$, pulsars cannot be white dwarfs and are identified with neutron stars.

S172. The gravitational force on a mass m is

$$F_p = \frac{GM_p m}{R_p^2},$$

on the planet, and

$$F_e = \frac{GM_e m}{R_e^2}$$

on the Earth. Since the mass's weight on the planet is half that on the Earth we have $F_p = F_e/2$, so the two equations give

$$\frac{GM_p m}{R_p^2} = \frac{GM_e m}{2R_e^2}.$$

With $M_p = 2M_e$ as given, we find $R_p^2 = 4R_e^2$, i.e. $R_p = 2R_e$.

S173. By the same argument as in S168 we have the relation $a^3 = GM_\odot P^2/(4\pi^2)$ between a and the year $P = 3 \times 10^7$ s. Using the value of G we find $M_\odot = 4\pi^2 a^3/(GP^2) = 2 \times 10^{30}$ kg.

In practice, the relative weakness of terrestrial gravitation means that the combination $GM_\odot$ is much better known than G or $M_\odot$ separately.

S174. Since the Earth is approximately spherical

$$g = \frac{GM_e}{R_e^2},$$

so $M_e = gR_e^2/G = 6 \times 10^{24}$ kg. This estimate is quite accurate.

S175. By conservation of energy we have

$$E_2 = E_1,$$

where E_2, E_1 are the total energies of the gun–bullet system at the top of the bullet's flight and immediately before firing. The former is the gravitational potential energy $U_g = mgh$, while the latter is the elastic potential energy $U_{\text{el}} = kx^2/2$, where k is the spring constant when the spring is compressed by an amount x. Thus

$$mgh = U_{\text{el}}.$$

The elastic potential energy is the same on the Earth and the Moon, while the gravitational potential energy has the values $mg_e h_e, mg_m h_m$ in the two cases, where m is the bullet's mass and g_e, g_m are the accelerations due to gravity on the Earth and the Moon respectively.

$$mg_e h_e = mg_m h_m,$$

so that $g_m = (h_e/h_m)g_e = g/6 = 1.63 \, \text{m s}^{-2}$.

S176. Let the satellite be at a radial distance R from the Earth's center. The resultant force ΣF_x on the satellite is the Earth's gravitational force

$$F_g = \frac{GM_e m}{R^2},$$

where m is the mass of the satellite. This must equal the centripetal force $m\omega^2 R$ required to keep the satellite in uniform circular motion with angular velocity $\omega = 2\pi/24 \; \mathrm{rad\,h^{-1}} = 7.27 \times 10^{-5} \; \mathrm{rad\,s^{-1}}$, i.e.

$$\frac{GM_e m}{R^2} = m\omega^2 R,$$

so that we require $R = (GM_e/\omega^2)^{1/3}$. Inserting the values of M_e and ω we find $R = 4.24 \times 10^7$ m. Subtracting the Earth's radius R_e, we find the height of the satellite as $h = R - R_e = 3.60 \times 10^7$ m.

This large value (almost $6R_e$) explains the high cost of launching such satellites. Because they remain fixed over the Earth they are nevertheless indispensable for communications, etc.

S177. The satellite must orbit the center of the Earth. A geostationary satellite over a point not on the equator would not do this.

S178. The shuttle's orbit has radius $a = R_e + H$, where R_e is the Earth's radius. If its velocity and mass are M, v, the gravitational and centripetal forces on it (see e.g. S176) are

$$F_g = \frac{GM_e M}{a^2}, \quad F_c = \frac{Mv^2}{a}$$

where M_e is the Earth's mass. These forces are in balance as the shuttle is in a circular orbit, so $v^2 = GM_e/a$. The satellite (mass m) has the same angular velocity $\omega = v/a = (GM_2/a^3)^{1/2}$, but is held at a radius $a + h$, so the corresponding forces on it are

$$f_g = \frac{GM_e m}{(a+h)^2}, \quad f_c = m\omega^2(a+h) = \frac{GM_e m}{a^3}(a+h) > f_g.$$

Gravity is therefore unable to supply the required centripetal force to keep the satellite in an orbit of radius $a + h$, and the initial motion is outwards, i.e. away from the shuttle and the Earth. (The satellite will go into a slightly elliptical orbit.)

S179. The retro rocket gives forward momentum to its exhaust gases. Since the shuttle and rocket are a closed system, momentum is conserved and this must slow the shuttle slightly. Gravity will now be larger than the centripetal force needed to hold the shuttle in its original orbit, and it will fall to a lower altitude (in fact its orbit will become elliptical, as for the satellite in the previous question). This is the basic method for bringing the shuttle back to Earth.

S180. The gravitational acceleration $F = GM/r^2$ of the satellite must supply its centripetal acceleration v^2/r. Equating, we find $v = (GM/r)^{1/2}$. The angular momentum per unit mass is $h = rv = (GMr)^{1/2}$. The atmospheric drag exerts a torque on the satellite's motion, which reduces its angular momentum per unit mass h. Since $h \propto r^{1/2} \propto 1/v$, this actually speeds the satellite up. This occurs because the satellite goes into an orbit at smaller r. It is a general property of gravitating systems that a loss of total (kinetic plus potential) energy always leads to an *increase* of kinetic energy, while the potential energy becomes more negative.

S181. The ratio of the Sun's pull to the Earth's is $M_\odot r^2/M_e a^2 = 2.3$. In fact both the Earth and the Moon are in nearly circular orbits about the Sun. They perturb each other's orbits – viewed from the Sun, the Moon performs a tiny "rosette" about the Earth's orbit (see Figure). The Moon cannot leave its orbit (and us) because of its angular momentum about the Sun.

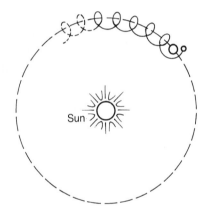

S182. A point on the planet's surface has to move in a circle about the Sun with angular velocity ω, so the effective gravity is $g_{\text{eff}} = N/m$, where N is the normal force exerted by the ground on a body of mass m. From the Figure we find

$$N + \frac{GMm}{(a-R)^2} - mg = m\omega^2(a-R),$$

for the point nearest to the Sun, so that

$$g_{\text{eff}} = g - \frac{GM}{(a-R)^2} + (a-R)\omega^2. \tag{1}$$

For the point furthest from the Sun we find

$$mg + \frac{GMm}{(a+R)^2} - N = m\omega^2(a+R),$$

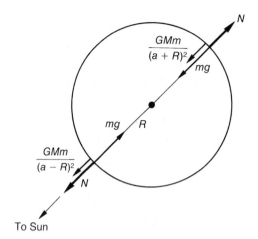

so that

$$g_{\text{eff}} = g + \frac{GM}{(a+R)^2} - (a+R)\omega^2, \tag{2}$$

where g is the usual surface gravity ($= Gm_p/R^2$, with m_p the planet's mass). Moreover, the gravitational pull of the Sun must supply the centripetal acceleration to keep the center of mass of the planet in orbit, i.e.

$$m_p a \omega^2 = \frac{GM}{a^2} m_p. \tag{3}$$

Now we can write

$$\frac{1}{(a-R)^2} = \frac{(a+R)^2}{(a^2-R^2)^2} = \frac{a^2 + 2aR + R^2}{a^4 + 2a^2R^2 + R^4} \approx \frac{a^2 + 2aR}{a^4} = \frac{1}{a^2} + \frac{2R}{a^3},$$

remembering that $a \gg R$ (if you are familiar with the binomial theorem, this result follows immediately). Similarly

$$\frac{1}{(a+R)^2} \approx \frac{1}{a^2} - \frac{2R}{a^3}.$$

Using these results in (1) and (2) and using $\omega^2 = GM/a^3$ from (3), we find

$$g_{\text{eff}} \approx g - 3\frac{GMR}{a^3}.$$

in both cases. On the circle equidistant from these points, the inward effective gravity is clearly just $g_{\text{eff}} = g$, as the Sun's gravity and the centrifugal acceleration just cancel. The effective gravity is thus highest here, and lowest at the nearest and furthest points.

S183. Hydrostatic equilibrium (see Summary in Ch. 3) requires that the water height h obeys

$$P = \rho g_{\text{eff}} h,$$

where ρ is the density of water and P is the pressure at the bottom of the ocean. If the ocean is static, P must be the same all over the planet, so

$$h \propto g_{\text{eff}}^{-1}.$$

The ocean is thus deepest ($h = d$) at the nearest and furthest points to the Sun, and shallowest ($h = s$) on the circle equidistant from them. The ratio of depths is

$$\frac{s}{d} = \left(g - 3\frac{GMR}{a^3} \right) g^{-1} = 1 - 3\frac{GMR}{a^3 g} = 1 - 3\frac{MR^3}{m_p a^3},$$

where we have used $g = Gm_p/R^2$, with m_p the planet's mass, in the last step. As the planet rotates, an observer on a small island would notice the ocean level rise and fall twice per revolution (i.e. twice per "day"), reaching its maximum height as the island passes through its nearest and furthest points from the Sun.

S184. From the last equation of the previous answer, the ratio of lunar and solar tides is $M_m a^3 / M_\odot b^3 = 2.15$. The tides are then highest when the Sun and Moon line up on either the same or opposite sides of the Earth, i.e. new moon or full moon. These are the so-called spring tides. The tides are lowest when the Sun and Moon pull at right angles at the Earth, and give the so-called neap tides. The answers given above predict the height of the tides on a planet completely covered by water, and give a value of order 0.5 m. Far from land, this is about the observed change in the height of the oceans. The tides observed near coasts can be much larger, as they result from water moving about in regions of varying depth in response to the change in g_{eff}.

S185. The Great Lakes and the Mediterranean are much smaller than the Earth's size, so g_{eff} is practically constant over them. They are almost landlocked, so as g_{eff} varies over the day their base pressures P simply vary in response, leaving their heights effectively unchanged, i.e. $P \propto g_{\text{eff}}$. This is impossible in the oceans as water flows to make P the same in regions with different g_{eff}.

S186. The angular momentum of the Moon is $L_m = M_m (GM_e b)^{1/2}$ (see S180). The Earth's spin angular momentum is $L_e = I\Omega$, where $\Omega = 2\pi/(\text{day})$ is its angular velocity in rad s^{-1} and I is the relevant moment of inertia. The angular momentum of the Earth–Moon system is conserved, so that $L_e + L_m = C$ or

$$I\Omega + M_m (GM_e b)^{1/2} = C, \tag{1}$$

where C is a constant. Since L_e decreases, L_m must increase, so b increases. Tidal dissipation will stop when the Earth spins synchronously with the

Moon, i.e. when $\Omega = (GM_e/b^3)^{1/2}$. The final value b_f of b is thus given by requiring

$$I\left(\frac{GM_e}{b_f^3}\right)^{1/2} + M_m(GM_e b_f)^{1/2} = C$$

With $b_f = 1.5b$, the length of the day becomes $P = 2\pi(b_f^3/GM_e)^{1/2} = 4.6 \times 10^6$ s, i.e. about 50 current days. (Direct evidence of this process is provided by determinations of the length of the day from studies of fossilized corals: 100 million years ago the day was only about 22 hours.)

S187. The potential energy of a mass m at distance r from the center of a gravitating mass M is $U(r) = -GMm/r$. If the mass starts from radius $r = R$ and gets to $r = \infty$, conservation of energy requires an initial speed $\geq v_{esc}$, where

$$\frac{1}{2}mv_{esc}^2 = 0 - \left(\frac{-GMm}{R}\right),$$

or

$$v_{esc} = \left(\frac{2GM}{R}\right)^{1/2}. \tag{1}$$

With $M = M_e, R = R_e$ the data given implies $v_{esc} = 11.2$ km s^{-1}.

S188. Using equation (1) of the previous solution, the escape speed for Saturn differs from that on Earth by a factor $(M_s/M_e)^{1/2}(R_e/R_s)^{1/2} = (95/9.4)^{1/2}$, so $v_{esc} = 11.2(95/9.4)^{1/2} = 35.6$ km s^{-1}.

S189. In a circular orbit at radius r the gravitational pull of the Earth must supply the centripetal force, so

$$\frac{mv^2}{r} = \frac{GM_e m}{r^2},$$

where m is the probe mass, i.e. $v^2 = GM_e/r$. The energy of the probe is thus

$$E = \frac{1}{2}mv^2 - \frac{GM_e m}{r} = -\frac{GM_e m}{2r}. \tag{1}$$

This must be the same as its energy at launch, i.e.

$$E = \frac{1}{2}mv_0^2 - \frac{GM_e m}{R_e} = -\frac{7GM_e m}{16R_e}, \tag{2}$$

where we have used $v_0 = (3/4)v_{esc} = (3/4)(2GM_e/R_e)^{1/2}$. Eliminating E between (1) and (2) we find $r = 8R_e/7$.

S190. Energy conservation requires

$$\frac{1}{2}mv_0^2 - \frac{GM_em}{R_e} = \frac{1}{2}mv^2 - \frac{GM_em}{r}. \tag{1}$$

where m is the rocket mass. Thus

$$v_0^2 - v^2 = \frac{2GM_e}{R_e}\left(1 - \frac{R_e}{r}\right) \tag{2}$$

At $r = 6R_e$ we have $v = v_0/10$, so (1) gives

$$\frac{99}{100}v_0^2 = \frac{5GM_e}{3R_e},$$

or $v_0 = 1.3(GM_e/R_e)^{1/2}$.

S191. The maximum height r is reached when $v = 0$, i.e. all kinetic energy has been converted into potential energy. Using energy conservation, expressed by equation (1) of the previous solution, with $v = 0$ and $v_0 = 1.3(GM_e/R_e)^{1/2}$, we find

$$\frac{GM_e}{r} = \frac{GM_e}{R_e} - \frac{1}{2}(1.3)^2\frac{GM_e}{R_e},$$

or

$$\frac{1}{r} = \frac{1}{R_e}(1 - 0.845),$$

giving $r = 6.45R_e$.

S192. The Earth's gravitational pull must supply the centripetal acceleration needed to keep the station in a circular orbit at any r, so

$$\frac{v^2}{r} = \frac{GM_e}{r^2}.$$

Thus with $r = 3R_e/2$ we find $v = (2GM_e/3R_e)^{1/2}$.

To achieve escape from $3R_e/2$ the minimum speed v_E with respect to the Earth must satisfy

$$\frac{1}{2}v_E^2 - \frac{GM_e}{3R_e/2} = 0,$$

by energy conservation. Thus $v_E = (4GM_e/3R_e)^{1/2}$. The most efficient way to arrange this is to use the speed the station already has. Then only a speed $v_r = v_E - v = 0.34(GM_e/R_e)^{1/2}$ is needed. The rocket is fired in the direction of the station's motion.

S193. Using equation (1) of S187 gives

$$R = \frac{2GM}{c^2}.$$

With the data given we find $R = 3$ km and 9 km for the $1M_\odot$ and $3M_\odot$ black holes respectively. In reality we need to use the General Theory of Relativity to evaluate R. However, the calculation given here, first performed by Laplace and Michell at the end of the 18th century, gives essentially the correct answer.

S194. Using the previous solution, the average density is

$$\rho = \frac{3M}{4\pi R^3} = 2 \times 10^{18} \text{ kg m}^{-3}$$

with the data given, so the densities are comparable. A *neutron star* has $R = 10$ km with $M = M_\odot$, and so also has nuclear density; the nucleons of its matter are as tightly packed as in an atomic nucleus.

S195. Since for black holes $R \propto M$, the average density found in the previous solution can be rewritten as

$$\rho = \frac{1.8 \times 10^{19}}{(M/M_\odot)^2} \text{ kg m}^{-3},$$

so with $M/M_\odot = 3 \times 10^9$ we get $\rho = 2$ kg m^{-3}, i.e. less than twice the density of air. Black holes are not necessarily very dense!

☐ RIGID BODY MOTION

S196. Using $v = \omega R$, we get $\omega = v/R = 10/0.5 = 20$ rad s^{-1}. As the acceleration is uniform, we have $\omega = \omega_0 + \alpha t$, so that $\alpha = (\omega - \omega_0)/t$. With $\omega_0 = 0, t = 10$ s and ω as above, we find $\alpha = 2$ rad s^{-2}.

S197. The moment of inertia is

$$I = \Sigma mr^2 = m_1 R^2 + m_2 R^2 + 2m_3 R^2 = 9 \text{ kg m}^2.$$

Newton's second law applied to circular motion gives

$$\Gamma = I\alpha,$$

where Γ is the torque. In our case $\Gamma = RF$, so

$$\alpha = \frac{\Gamma}{I} = \frac{RF}{I} = \frac{1 \times 5}{9} = 0.56 \text{ rad s}^{-2}.$$

S198. From the definition,

$$I_x = 2m_3 R^2 = 6 \text{ kg m}^2,$$

$$I_y = m_1 R^2 + m_2 R^2 = 3 \text{ kg m}^2.$$

Note that $I_x + I_y = I$ from the previous question (perpendicular axes theorem).

S199. Using conservation of energy for either the rolling cylinder or sphere, starting from rest, we have

$$U = \frac{1}{2}mv^2 + \frac{1}{2}I\omega^2,$$

where $U = mgh$ is the potential energy difference, v is the linear velocity of the center of mass at the bottom of the plane, ω is the angular velocity, and I the moment of inertia about the rolling axis. Because the bodies roll rather than slipping we must have $\omega = v/r$. Thus

$$2mgh = \left(m + \frac{I}{r^2} \right) v^2,$$

or

$$v^2 = \frac{2mgh}{(m + I/r^2)}.$$

For the cylinder, $I = mr^2/2$ (independent of its length). Thus $v_c^2 = 2mgh/(3m/2) = 4gh/3$ or $v_c = 2(hg/3)^{1/2}$. This is the final velocity of the center of mass along the slope. The time t_c is given by the kinematic formula $v = at$ as $t_c = v_c/a$ with $a = g \sin \alpha = g\sqrt{3}/2$, Thus

$$t_c = \frac{v_c}{g \sin \alpha} = \frac{4}{3g}(gh)^{1/2} = 1.33 \left(\frac{h}{g} \right)^{1/2}.$$

Similarly for the sphere, which has $I = 2mr^2/5$, we find $v_s^2 = 2mgh/(7m/5) = 10gh/7$, or $v_s = (10gh/7)^{1/2}$. Then as before

$$t_s = \frac{v_s}{g \sin \alpha} = \left(\frac{40h}{21g} \right)^{1/2} = 1.38 \left(\frac{h}{g} \right)^{1/2}.$$

Note that since the moment of inertia of the sphere is smaller than that of the cylinder, the linear velocity of its center of mass is greater after the descent (less of the potential energy used to spin it up). Since the acceleration is the same in the two cases, it takes a *longer* time to accelerate the sphere. By extending this argument we see that when there is no rolling ($\omega = 0$), the time t_0 must be longer still. Thus here $v_c = v_s = (2gh)^{1/2}$ and

$$t_0 = \frac{(2gh)^{1/2}}{g \sin \alpha} = \left(\frac{8h}{3g} \right)^{1/2} = 1.63 \left(\frac{h}{g} \right)^{1/2}.$$

S200. To complete the circular loop the cylinder's velocity v at the top of the loop must be $\geq$ the value v_c at which the centripetal force mv_c^2/R is exactly supplied by the gravity force mg, i.e. $v_c = (gR)^{1/2}$. We can find v by using conservation of energy: the total mechanical energy (kinetic + potential) when the cylinder is released must equal the value at the top of the loop, i.e.

$$mgh = \frac{1}{2}mv^2 + \frac{1}{2}I\omega^2 + 2mgR,$$

where I is the cylinder's moment of inertia and its angular velocity ω equals v/r because the cylinder does not slip. Thus

$$h = \frac{v^2}{2g}\left(1 + \frac{I}{mr^2}\right) + 2R. \tag{1}$$

This takes the value $h = h_m$ when $v = v_c$. Using $I = mr^2/2$ for the solid cylinder we get $h_m = 3v_c^2/4g + 2R$, and substituting for v_c we find $h_m = 3R/4 + 2R = 2.75R$.

For a hollow cylinder all that changes is the moment of inertia, i.e. $I = mr^2$. Using this in (1) with $h = h_m = 2.75R$ we can find the new value of v, i.e. $2.75R = (v^2/2g) \times 2 + 2R$, or $v^2 = 0.75gR < v_c^2 = gR$. Thus $v < v_c$ and the hollow cylinder will leave the loop.

201. This result can be proved directly from the definition, but a simple way of obtaining it uses the fact that the energy of the body must be the same regardless of whether we think of it as:

(a) – rotating with angular velocity ω about an axis through its center of mass, which is itself moving with linear velocity $v = d\omega$ with respect to the new axis, or

(b) – simply rotating about this new axis with angular velocity ω.

In case (a) its total energy is $\frac{1}{2}I\omega^2 + \frac{1}{2}Mv^2 = \frac{1}{2}(I + Md^2)\omega^2$, while in case (b) the total energy is $\frac{1}{2}I'\omega^2$, where I' is the new moment of inertia. Equating the two expressions we see that $I' = I + Md^2$.

S202. Newton's second law for the linear motion of the mass gives (see Figure)

$$mg - T = ma \tag{1}$$

For the angular motion of the pulley we have similarly

$$RT = I\alpha,$$

where $I = MR^2/2$ is the pulley's moment of inertia and α its angular acceleration. The no-slip condition is $\alpha = a/R$. Thus $RT = (MR^2/2)(a/R)$ so that $T = Ma/2 = ma$. Substituting in (1) we have $mg = 2ma$ or $a = g/2$ and $T = mg/2$.

The angular momentum after the mass descends a height R is $L = I\omega + mvR$, where v is the mass's velocity and ω the corresponding angular velocity of the pulley at this time. Since $\omega = v/R$ (no-slip) and $I = MR^2/2 = mR^2$ as before, we have $L = 2mvR$. The velocity follows from the kinematic formula $v^2 = v_0^2 + 2ax$ with $v_0 = 0$, $a = g/2$ and $x = R$. This gives $v = (gR)^{1/2}$ and thus $L = 2m(gR^3)^{1/2}$.

S203. By conservation of angular momentum the top must maintain the value L in the vertical direction. If it is pushed through an angle θ away from the vertical, this component of angular momentum becomes $L\cos\theta$, so the deficit $(1 - \cos\theta)L$ has to be made up somehow. The top achieves this by *precessing*, i.e. it rotates its spin axis around the vertical. This rotational motion returns the missing vertical component of angular momentum. Once L is reduced to a small value (the angular momentum is gradually transferred to the surface on which the top rests, through friction at the point), the precession angle θ gets so large that the sides of the top hit the surface and it falls over.

S204. The bullet acquires angular momentum about an axis parallel to the barrel. Because angular momentum is conserved (see previous solution) this keeps the bullet pointing stably in this direction and so improves accuracy.

S205. Conservation of angular momentum gives

$$I\omega = I\omega' + mr^2\omega',$$

where ω' is the new angular velocity of the turntable + glue. Thus $\omega' = I\omega/(I + mr^2)$. With $I = MR^2/2, m = M/10$ and $r = 3R/4$, we find

$$\omega' = \frac{\frac{1}{2}MR^2}{\frac{1}{2}MR^2 + \frac{1}{10}M(0.75)^2R^2}\omega = 0.9\omega.$$

S206. A pendulum of moment of inertia I and mass M has period

$$P = 2\pi \left(\frac{I}{MgL_{CM}} \right)^{1/2},$$

where L_{CM} is the distance of the center of mass from the pivot.

(a) Here $I = Ml^2/12 + Ml^2/4 = Ml^2/3$ (parallel axes theorem) and $L_{CM} = l/2$. Thus $P = 2\pi(2l/3g)^{1/2} = 1.16$ s.

(b) The moment of inertia here is given by $I = I_{CM} + MR^2$, where R is the distance of the pivot from the center of mass (parallel axes theorem). Thus

$$I = \frac{Ml^2}{12} + M \left(\frac{l}{2} - l_C \right)^2,$$

since $I_{CM} = Ml^2/12$ for a uniform rod. Moreover $L_{CM} = R = l/2 - l_C$. With $l_C = l/4$ we have

$$I = \frac{Ml^2}{12} + M \left(\frac{l}{4} \right)^2 = 0.146Ml^2$$

and $L_{CM} = l/4$. Thus $P = 2\pi(0.146l/0.25g)^{1/2} = 1.08$ s.

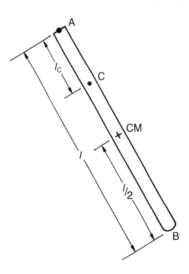

S207. By the parallel axes theorem (see P201) the moment of inertia of an extended arm about the skater's axis is $mL^2/12 + m(L/2 + R)^2$. If the arms are by the skater's side, the moment of inertia is just mR^2. Thus the moments of inertia before and after he drops his arms are

$$I_b = \frac{MR^2}{2} + 2 \left[\frac{mL^2}{12} + m(\frac{L}{2} + R)^2 \right],$$

$$I_a = \frac{MR^2}{2} + 2mR^2.$$

As no external torques act on the skater, angular momentum is conserved, i.e.

$$I_b\omega_b = I_a\omega_a,$$

where ω_a is the angular velocity after he drops his arms. With the data given, we find $I_b = 4.49$ kg m^2, $I_a = 1.76$ kg m^2. Hence the angular velocity after the skater drops his arms is $\omega_a = (I_b/I_a)\omega_b = 15.3$ rad s^{-1}.

S208. As soon as the man starts to walk, the disc will begin to rotate with angular velocity ω_d. Conservation of angular momentum gives

$$0 = I\omega_d + mrv,$$

where $I = Mr^2/2$ is the disc's moment of inertia. Thus $\omega_d = -mrv/I = -2(m/M)(v/r)$, i.e. the disc rotates in the opposite direction. The man's angular velocity with respect to the Earth is $\omega = v/r$, so his angular velocity with respect to the disc is $\omega' = \omega - \omega_d = (1 + 2m/M)(v/r)$. Thus $\omega' = 2(2/2.5) = 1.6$ rad s^{-1}, and the period (the time to return to the same point) on the disc is $P = 2\pi/\omega' = 2\pi/1.6 = 3.93$ s. If the man stops, the disc must also stop, as the total angular momentum must remain zero.

S209. By Newton's second law, the equation of linear motion is

$$-mg\mu = ma,$$

and the corresponding relation for the angular motion is

$$mg\mu R = \tfrac{2}{5}mR^2\alpha,$$

where α is the angular acceleration, since the moment of inertia of the pool ball about an axis through its center is $2mR^2/5$. From the first equation $a = -\mu g$, and thus $v(t) = v_0 - g\mu t$ by the usual kinematic formula. Similarly $\alpha = 5\mu g/2R$ and $\omega(t) = 5\mu gt/2R$, since the ball starts with zero angular velocity. The condition for pure rolling motion is $v = \omega R$, or

$$v_0 - g\mu t = \frac{5}{2}\mu gt.$$

Solving for t we find that the ball starts pure rolling after a time $t = 2v_0/(7\mu g)$. The velocity is then $v = v_0 - g\mu t = v_0 - 2v_0/7 = 5v_0/7$.

210. If the ball gives the bat linear momentum p, it also gives it angular momentum $p(x - l)$ about its center of mass, so two things happen:

    (a) – the bat's center of mass acquires velocity $v = p/M$, and

    (b) – the bat rotates about its center of mass with angular velocity $\omega = (x - l)p/I$.

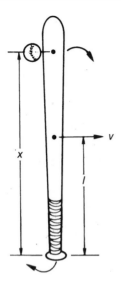

If $x > l$, this rotation produces a linear velocity at the bat's base in the opposite direction from v (see Figure). The reaction force at the player's hands vanishes if the total velocity there is zero, i.e. the bat pivots about the player's hands. The condition for this is $v = l\omega$, i.e. $p/M = l(x-l)p/I$, or

$$h = l + \frac{I}{Ml}.$$

If the bat is regarded as a uniform rod of length $2l$, the appropriate value of I is $I = Ml^2/3$, so $x = 4l/3$, i.e. the player should aim to strike the ball about two-thirds of the length of the bat from the handle. This is the so-called *center of percussion* or "sweet spot." An impact here gives the feeling of hitting the ball "off the meat," i.e. without jarring the hands.

S211. This is actually exactly the same physical problem as studied in the previous question. Here the point where the ball rests on the table plays the role of the baseball player's hands. The condition that the ball should initially pivot about this point is

$$h = l + \frac{I}{Ml}$$

as before. With $I = 2Ml^2/5$ for a sphere, we find $h = 7l/5$, i.e. the player should cue the ball 7/10 of a diameter above the table. The cushions on a pool table are at this height so that a rolling ball rebounds without skidding.

S212. If there is friction at the disk axis, angular momentum is lost by the disk to the Earth. When the man stops walking, the disk's angular momentum is now too small to cancel his angular momentum completely, so he and the disk rotate slowly in his forward direction.

CHAPTER **TWO**

ELECTRICITY AND MAGNETISM

□ ELECTRIC FORCES AND FIELDS

S213. As Q has the same sign as q_1, q_2 the forces on it are both repulsive. Thus taking the direction from q_1 to q_2 as positive (see Figure) the net force on Q is

$$F = F_1 - F_2 = \frac{q_1 Q}{4\pi\epsilon_0 x^2} - \frac{q_2 Q}{4\pi\epsilon_0 (d-x)^2} = \frac{Q}{4\pi\epsilon_0}\left[\frac{q_1}{x^2} - \frac{q_2}{(d-x)^2}\right], \quad (1)$$

where x is the distance of Q from q_1.

With $x = d/2$ we find

$$F = \frac{4Q}{4\pi\epsilon_0 d^2}(q_1 - q_2) = 4 \times 9 \times 10^9 \times 10^{-5} \times (-2 \times 10^{-5}) = -7.2 \text{ N}.$$

The force acts on Q in the direction of q_1. We can find the point where the force vanishes by setting $F = 0$ in equation (1). Thus

$$\frac{q_1}{x^2} - \frac{q_2}{(d-x)^2} = 0,$$

or

$$\left(\frac{q_2}{q_1}\right)^{1/2} = \frac{d-x}{x} = \frac{d}{x} - 1$$

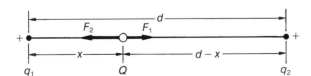

i.e.

$$x = \frac{d}{1 + (q_2/q_1)^{1/2}} . \tag{1}$$

Substituting we find $x = 1/(1 + 2^{1/2}) = 0.414$ m.

S214. The total forces F_1, F_2 on q_1, q_2 should vanish, i.e.

$$F_1 = \frac{1}{4\pi\epsilon_0} \left(\frac{q_1 q_2}{l^2} + \frac{q_1 Q}{x^2} \right) = 0, \tag{1}$$

$$F_2 = \frac{1}{4\pi\epsilon_0} \left(\frac{q_1 q_2}{l^2} + \frac{q_2 Q}{(l-x)^2} \right) = 0.$$

Eliminating $q_1 q_2/l^2$ gives

$$\frac{q_1}{x^2} = \frac{q_2}{(l-x)^2},$$

so that

$$\frac{x}{l-x} = \left(\frac{q_1}{q_2} \right)^{1/2} = 3$$

and $x = 3l/4 = 0.75$ m. Using (1) we get $Q = -q_2(x^2/l^2) = -5.6 \times 10^{-3}$ C.

S215. The resultant force F is the sum of the electrostatic forces F_1, F_2 exerted by each charge. We must add these forces component by component, so that

$$F_x = F_{1x} + F_{2x},$$

$$F_y = F_{1y} + F_{2y}.$$

Now $F_{1x} = 0$ (force only along the y-axis) and similarly $F_{2y} = 0$. Thus

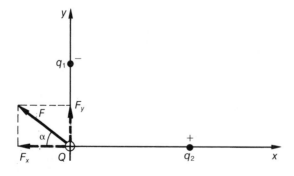

$$F_x = F_{2x} = \frac{-1}{4\pi\epsilon_0} \frac{Qq_2}{x_2^2} = \frac{-9 \times 10^9 \times 10^{-6}}{16} = -563 \text{ N (repulsion)},$$

$$F_y = F_{1y} = \frac{-1}{4\pi\epsilon_0} \frac{Qq_1}{y_1^2} = \frac{9 \times 10^9 \times 0.5 \times 10^{-6}}{9} = 500 \text{ N (attraction)}.$$

Thus $F = (500^2 + 563^2)^{1/2} = 753$ N. From the Figure this force makes an angle α to the negative x-axis, where $\tan\alpha = |F_y|/|F_x|$, i.e. $\alpha = 41.6°$.

S216. See the Figure. The first charge gives a force F_x along the x-axis:

$$F_x = \frac{q_1 q_3}{4\pi\epsilon_0 x_1^2} = 9 \times 10^9 \frac{-2 \times 10^{-6} \times 10^{-6}}{(0.08)^2} = -2.813 \text{ N},$$

while the second charge gives a force along the y-axis:

$$F_y = \frac{q_2 q_3}{4\pi\epsilon_0 y_2^2} = 9 \times 10^9 \frac{3 \times 10^{-6} \times 10^{-6}}{(0.1)^2} = 2.7 \text{ N}.$$

The total force is therefore $F = (F_x^2 + F_y^2)^{1/2} = 3.9$ N, acting at an angle $\alpha = \tan^{-1}|F_y/F_x| = 43.83°$ to the x-axis in the negative-x, positive-y direction.

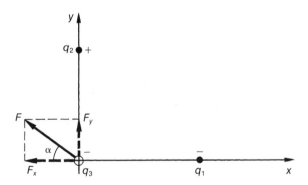

S217. This is essentially the same as the second part of S213, since the forces on the sliding sphere are opposed whatever the sign of Q. Substituting $d = l, q_2/q_1 = 4$ into equation (1) of S213 shows that

$$x = \frac{l}{1 + 4^{1/2}} = \frac{l}{3},$$

i.e. the sliding sphere will be in equilibrium at distance $l/3$ from the smaller charge q_1.

S218. The diagonals of the square cross at right angles, so we take them as the axes of a coordinate system with origin at the center (see Figure). The field **E** at the center is the sum of the fields produced by each charge. The latter are directed radially about each charge, with strength $q/(4\pi\epsilon_0 d^2)$, where d, the distance of each charge from the center, is half of the diagonal length, i.e. $d = a\sqrt{2}/2$. Since the fields are radial, the x and y components of **E** are

$$E_x = E_2 + E_3 = \frac{1}{4\pi\epsilon_0}\left(-\frac{Q}{d^2} + \frac{Q}{d^2}\right) = 0,$$

$$E_y = E_1 + E_4 = \frac{-1}{4\pi\epsilon_0}\left(\frac{Q}{d^2} + \frac{Q}{d^2}\right) = \frac{-Q}{\pi\epsilon_0 a^2}.$$

Substituting $Q = 1\,\mathrm{C}$, etc. we find a field $E = 9 \times 10^9$ N/C in the $-y$-direction, i.e. towards the charge $-Q$.

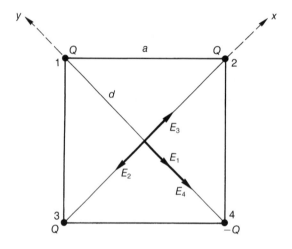

S219. The proton has charge $q = +e = 1.6 \times 10^{-19}\,\mathrm{C}$, so the electric field is $E = q/(4\pi\epsilon_0 a^2) = 5.17 \times 10^{11}\,\mathrm{N\,C^{-1}}$, and the force on the electron is $F = eE = 8.26 \times 10^{-8}$ N inwards. In circular motion this must supply the centripetal force $F = m_e v_e^2/a$, where m_e, v_e are the electron's mass and velocity. Hence $v_e = (aF/m_e)^{1/2} = 2.20 \times 10^6\,\mathrm{m\,s^{-1}}$ and the period is $P = 2\pi a/v_e = 2\pi(m_e a/F)^{1/2} = 1.51 \times 10^{-16}$ s.

S220. By Gauss's law the charge and field are connected by $Q_e = 4\pi\epsilon_0 R_e^2 E_e = 5.92 \times 10^5$ C.

S221. Vertical force balance requires $qE = mg$ or $q = mg/E$. Substituting $m = 0.01$ kg and $E = E_e = 130\,\mathrm{N\,C^{-1}}$ gives $q = 9.8 \times 0.01/130 = 7.54 \times 10^{-4}$ C.

S222. As the first two charges have the same sign, the charge Q must lie on the line joining them, as otherwise the component of force on Q towards that line does not vanish. To overcome the electrostatic repulsion between the original pair of charges, Q must clearly have the opposite sign and lie between them. Let its distance from charge q be x (see Figure). Then the vanishing of the electrostatic forces on the charges $q, 9q$ and Q gives us the three equations

$$\frac{qQ}{x^2} + \frac{9q^2}{l^2} = 0 \tag{1}$$

$$\frac{9q^2}{l^2} + \frac{9qQ}{(l-x)^2} = 0 \tag{2}$$

$$\frac{9qQ}{(l-x)^2} - \frac{qQ}{x^2} = 0. \tag{3}$$

We note that (3) is automatically satisfied if (1, 2) hold, as can be seen by subtracting (1) from (2). From (3) we get $9x^2 = (l-x)^2$, or taking the square root of each side, $l - x = \pm 3x$. This leads to $x = l/4, -l/2$. Only the first root is physical, the second spurious root being introduced by the operation of taking the square root above. With $x = l/4$ we now find from (1) that $Q = -9x^2q/l^2 = -9q/16$. As expected, Q turns out to be negative.

S223. The force on the electron is $F = eE_0$ *upwards* (as the electron's charge is negative). Thus its acceleration is $a = eE_0/m_e = 1.76 \times 10^{14}$ m s^{-2}. This is far larger than $g = 9.8$ m s^{-2}, so the neglect of gravity is justified. The horizontal motion is uniform, so time of flight between the plates is $t = l_0/v_0 = 10l_0/c$, and the deflection is $y = at^2/2 = 50al_0^2/c^2 = 0.098$ m upwards.

S224. No horizontal forces act on the electrons in the beam, so at time t after injection they are at horizontal distance $x = v_e t$. In the vertical direction gravity is negligible in comparison with the Coulomb force $-eE_0$, which produces a constant acceleration $-eE_0/m_e$. The vertical displacement at time t is thus $y = -eE_0 t^2/2m_e$. Eliminating t we find the path

$$y(x) = -\frac{eE_0}{2m_e v_e^2} x^2.$$

(a) Reversing the field raises the beam symmetrically, so it hits the screen at 10 cm above the horizontal.

(b) At $x = l$ we have $y = -h (= -10$ cm$)$, so substituting in the equation above we find $l = (2m_e h/eE_0)^{1/2} v_e = 2.4$ cm.

S225. The electrons acquire horizontal velocity v_x given by energy conservation:

$$\tfrac{1}{2} m_e v_x^2 = |-eV|,$$

i.e. $v_x = (2eV/m_e)^{1/2}$. The potential difference V_P between the plates gives an electric field $E = V_P/d$, which deflects the electrons. This implies constant vertical acceleration $a = eE/m_e = eV_P/(m_e d)$: the electrons spend a time $t = l/v_x = l(m_e/2eV)^{1/2}$ passing between the plates, so using the kinematic formula $y = y_0 + at^2/2$ the deflection is

$$y = \frac{eV_P}{m_e d} \frac{t^2}{2} = \frac{eV_P}{m_e d} l^2 \frac{m_e}{4eV} = \frac{l^2}{4d} \frac{V_P}{V}.$$

The maximum deflection which still allows the electrons to miss the plates is $y = d/2$, and this requires $V_P = 2(d/l)^2 V$.

S226. Let the balls have charges q_1, q_2. Then vertical force balance requires $-E_1 q_1 = mg, -E_2 q_2 = mg$, where $m = 4\pi r^3 \rho/3$ is the ball's mass. Using $\rho = 0.8$ g cm^{-3} $= 800$ kg m^{-3} we find $m = 3.35 \times 10^{-15}$ kg, and hence $q_1 = -3.26 \times 10^{-19}$ C, $q_2 = -4.89 \times 10^{-19}$ C. Hence $-e = -(q_1 - q_2) = 1.59 \times 10^{-19}$ C. Note that in reality we cannot be sure that the charges differ by exactly $-e$, rather than some multiple of it. In practice the experimenter looks to find the smallest charge difference; all other differences should be integer multiples of this one.

S227. From the Figure, we have for each mass

$$T \sin \theta = F_e,$$

$$T \cos \theta = mg$$

so that the electrostatic repulsive force is

$$F_e = mg \tan \theta. \tag{1}$$

But we have

$$F_e = \frac{1}{4\pi\epsilon_0} \frac{Q^2}{d^2}$$

and

$$d = 2l_0 \sin \theta.$$

Thus

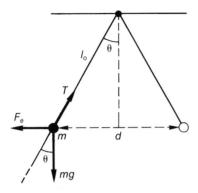

$$F_e = \frac{1}{4\pi\epsilon_0} \frac{Q^2}{4l_0^2 \sin^2\theta}.$$

Using this in (1) we find

$$Q^2 = (4\pi\epsilon_0).4l_0^2 mg \sin^2\theta \tan\theta$$

so $Q = 2.5 \times 10^{-5}$ C.

S228. If the number of electrons added to each ball is N, the charge on each is $q = -Ne$. The electrostatic repulsion is then $F = (Ne)^2/(4\pi\epsilon_0)d^2 = 2.3 \times 10^{-28} N^2$ N. This reaches 1000 N for $N^2 = 4.34 \times 10^{30}$ or $N = 2.08 \times 10^{15}$. The typical number of electrons in 10 g of metal is much larger, of order Avogadro's number $\sim 10^{24}$.

S229. The electrostatic and gravitational forces are $F_e = q_\alpha^2/4\pi\epsilon_0 d^2 = 2.3$ N and $F_g = -Gm_\alpha^2/d^2$. Thus $|F_e/F_g| = q_\alpha^2/(4\pi\epsilon_0 Gm_\alpha^2) = 3.1 \times 10^{35}$. The electrostatic force is enormously larger than the gravitational force: note that the ratio is independent of distance since both obey inverse square laws ($F \propto d^{-2}$).

S230. We require $eE_0 = m_e g$, i.e. $E_0 = m_e g/e = 5.57 \times 10^{-11}$ N C^{-1}. The proton produces a field $E_p = e/(4\pi\epsilon_0 a_0^2) = 1.44 \times 10^{11}$ N C^{-1} at distance a_0. This is enormously larger, showing that gravity has a negligible effect on atomic structure.

S231. We can regard the cylinder as infinitely long. Consider a cylindrical surface of radius r and length l centered on the axis (see Figure). By symmetry the electric field must be radial, and so is perpendicular to the curved surface of the cylinder. Then by Gauss's law we have

$$\Phi = \frac{1}{\epsilon_0} Q_{\text{in}}$$

where the flux $\Phi = 2\pi r l E(r)$, and the charge enclosed by the surface is $Q_{\text{in}} = \pi r^2 l \rho_0$. Thus

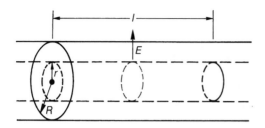

$$2\pi r l E(r) = \frac{1}{\epsilon_0}\pi r^2 l \rho_0,$$

or

$$E(r) = \frac{\rho_0 R}{2\epsilon_0}\frac{r}{R} = 5.66 \times 10^6 \frac{r}{R} \text{ N C}^{-1}.$$

S232. From the Figure we have the force components

$$F_x = 0, \text{ (by symmetry)}$$

$$F_y = -2\frac{1}{4\pi\epsilon_0}\frac{q^2}{2d^2}\cos\alpha = \frac{-1}{4\pi\epsilon_0}\frac{q^2}{(a^2+y^2)}\frac{y}{\sqrt{a^2+y^2}} = \frac{-1}{4\pi\epsilon_0}\frac{q^2 y}{(a^2+y^2)^{3/2}}.$$

Thus for $y = 0$ we have $F_x = F_y = 0$, so that the origin is indeed an equilibrium point. For $y \ll a$ we can neglect the term y^2 in the denominator, so that

$$F_y = \frac{-q^2}{4\pi\epsilon_0 a^3}y.$$

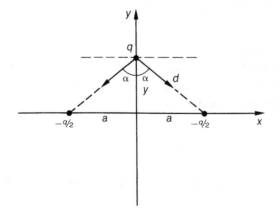

This is the equation for simple harmonic motion, with frequency ω given by dividing the coefficient of y by the mass m and taking the square root, i.e. $\omega = (q^2/4\pi\epsilon_0 a^3 m)^{1/2} = 335$ rad s^{-1}. Hence the period is $P = 2\pi/\omega = 0.019$ s.

S233. The electric field E of the line charge is radial. We apply Gauss's law to a cylinder of radius R and length l about the line charge. The flux of electric field is $\Phi = 2\pi R l E$ and must equal $1/\epsilon_0$ times the enclosed charge, $q = \lambda l$, so that $E = \lambda/(2\pi\epsilon_0 R)$. The resulting electrostatic force on the orbiting charge is qE, which acts radially inwards, as $\lambda < 0$. For this to supply the centripetal force, mv^2/R, requires $v^2 = -\lambda q/(2\pi\epsilon_0 m)$. Note that the radius of the orbit drops out. Inserting the values given shows that $v = 1.9 \times 10^4$ m s^{-1}.

S234. On the x-axis the field components are (see Figure)

$$E_x = \frac{1}{4\pi\epsilon_0}\left(\frac{q\cos\theta}{x^2+a^2} - \frac{q\cos\theta}{x^2+a^2}\right) = 0$$

$$E_y = \frac{1}{4\pi\epsilon_0}\left(\frac{-q\sin\theta}{x^2+a^2} - \frac{q\sin\theta}{x^2+a^2}\right) = -2\frac{1}{4\pi\epsilon_0}\frac{q}{x^2+a^2}\sin\theta.$$

Now $\sin\theta = a(x^2+a^2)^{-1/2}$, so

$$E_y = -\frac{1}{4\pi\epsilon_0}\frac{2qa}{(x^2+a^2)^{3/2}}.$$

Clearly, for $x \gg a$ we have $E_y \propto x^{-3}$.

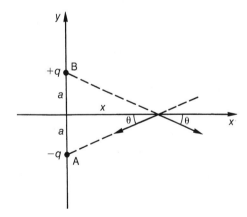

S235. We can regard the plate as infinite with uniform surface charge density $\sigma = 100Q/A = 100Q/(100d)^2 = 10^{-2}Q/d^2$. Then Gauss's law shows that the resulting electric field has components $E_x^{\text{plate}} = \sigma/(2\epsilon_0) = 5 \times 10^{-3}Q/(\epsilon_0 d^2)$; $E_y^{\text{plate}} = 0$. We must add to this the field of the shell. This is *zero* inside the shell, and equal to that of a point charge Q outside it (by Gauss's law). Hence inside the shell

$$E_x = E_x^{\text{plate}} = 5 \times 10^{-3} \frac{q}{\epsilon_0 d^2},$$

$$E_y = 0.$$

Outside the shell (see Figure) for any point $P(x, y)$

$$E_x = E_x^{\text{plate}} + E_x^{\text{shell}} = E_x^{\text{plate}} - E^{\text{shell}} \cos \alpha,$$

i.e.

$$E_x = 5 \times 10^{-3} \frac{Q}{\epsilon_0 d^2} - \frac{1}{4\pi\epsilon_0} \frac{Q}{(d-x)^2 + y^2} \frac{d-x}{[(d-x)^2 + y^2]^{1/2}}$$

and

$$E_y = E_y^{\text{shell}} = E^{\text{shell}} \sin \alpha = \frac{1}{4\pi\epsilon_0} \frac{Q}{(d-x)^2 + y^2} \frac{y}{[(d-x)^2 + y^2]^{1/2}}.$$

This is the general result for any point (x, y) outside the shell but not very close to the edges of the plate. Substituting $x = y = d/2$ for point P_2, we find

$$E_x = -0.107 \frac{Q}{\epsilon_0 d^2},$$

$$E_y = 0.113 \frac{Q}{\epsilon_0 d^2}.$$

The magnitude of the resultant field is thus

$$E = (0.107^2 + 0.113^2)^{1/2} \frac{Q}{\epsilon_0 d^2} = 0.156 \frac{Q}{\epsilon_0 d^2},$$

and it makes an angle θ with the negative x-direction, where

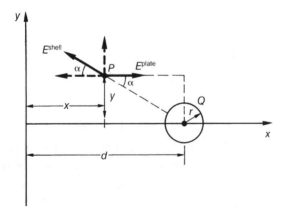

$$\tan\theta = \frac{0.113}{0.107} = 1.056,$$

or $\theta = 46.6°$.

S236. By spherical symmetry the electric field is radial everywhere. The value $E(r)$ of the field at radius r follows from applying Gauss's law in various regions (see Figure).

For $r < a$ we have

$$E(r) = \frac{q(r)}{4\pi\epsilon_0 r^2},$$

where $q(r)$ is the total enclosed charge, i.e. $q(r) = Qr^3/a^3$. Thus

$$E(r) = \frac{Qr}{4\pi\epsilon_0 a^3}.$$

For $a < r < 2a$ we have

$$E(r) = \frac{Q}{4\pi\epsilon_0 r^2},$$

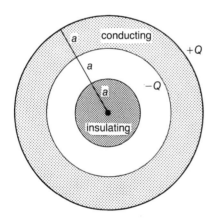

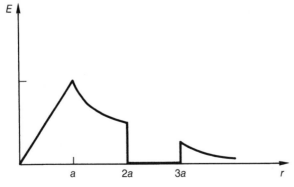

just as for a point charge. For $2a < r < 3a$ we have $E(r) = 0$, as this is the interior of a perfect conductor; a charge $-Q$ will be induced on the inside of the shell. For $r > 3a$ we have

$$E(r) = \frac{Q}{4\pi\epsilon_0 r^2},$$

as a charge $+Q$ is induced on the outside of the shell. See Figure for a graph of $E(r)$.

S237. $E(r)$ follows in each region (see Figure) by using Gauss's law. Inside the first sphere, a surface of constant r encloses total charge q, while between the two spheres the total enclosed charge is $-2q + q = -q$. Outside both spheres the enclosed charge is zero. Thus for $0 < r < R$ we have $E(r) = q/(4\pi\epsilon_0 r^2)$; for $R < r < 2R$ we have $E(r) = -q/(4\pi\epsilon_0 r^2)$; and for $r > 2R$ we have $E(r) = 0$.

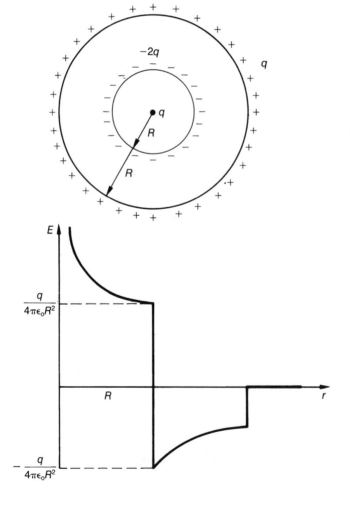

Note that $E(r)$ is discontinuous at each of the spheres (see Figure). This is characteristic of the effect of charge layers.

S238. By Gauss's law (see S233) the external electric field is $E_r = \lambda/(2\pi\epsilon_0 r)$, where λ is the total *linear* charge density (i.e. charge per unit length). Here $\lambda = \lambda_{\text{core}} + \lambda_{\text{sheath}}$, and unit length of the core and sheath have charges $\lambda_{\text{core}} = \rho\pi R^2$, $\lambda_{\text{sheath}} = 2\pi R\sigma$. To arrange that $E(r) = 0$ everywhere we must choose σ so that $\lambda = 0$, i.e. $\sigma = -\rho R/2$.

S239. By symmetry the field is directed radially outwards and depends only on r. Gauss's law applied to a cylinder of length L and radius $r \leq R$ gives

$$2\pi r L E = \frac{1}{\epsilon_0} \pi r^2 L \rho$$

so that $E(r) = \rho r/2\epsilon_0$.

For $r > R$ Gauss's law gives

$$2\pi r L E = \frac{1}{\epsilon_0} \pi R^2 L \rho$$

since the whole charge is included. Thus $E(r) = R^2 \rho/(2\epsilon_0 r)$. These results are sketched in the Figure.

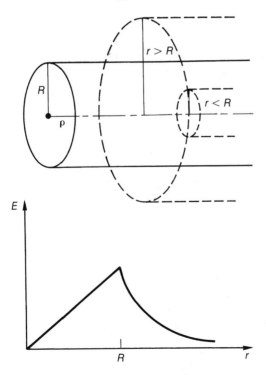

S240. We take the charge sheet as lying in the y, z plane and the point charge on the x-axis. The electric field of the charge sheet is uniform and has only an x-component:

$$E_x = \frac{\sigma}{2\epsilon_0}, \quad x > 0,$$

$$E_x = -\frac{\sigma}{2\epsilon_0}, \quad x < 0$$

by Gauss's law. The acceleration $a = qE_x/m$ is therefore also uniform and directed towards the charge sheet, since $\sigma < 0$. It is discontinuous at $x = 0$, i.e.

$$a = -\frac{q^2}{2\epsilon_0 md^2}, \quad x > 0,$$

$$a = \frac{q^2}{2\epsilon_0 md^2}, \quad x < 0,$$

where we recall that $\sigma = -q/d^2$. The charge accelerates from rest towards the layer; using the kinematic formula $v^2 = v_0^2 + 2as$, we find $v = (2|a|s)^{1/2}$, where s is the distance traveled from rest. Thus for $0 < x < d$ we have

$$v(x) = -\frac{q}{d}\left(\frac{d-x}{\epsilon_0 m}\right)^{1/2}.$$

The motion for $-d < x < 0$ is the mirror image of this. At $x = \pm d$ the charge halts momentarily before accelerating back towards the layer. The motion is clearly periodic with period $P = 4t$, where t is the time to reach the layer. From the kinematic formula $s = v_0 t + at^2/2$ the latter is $t = (2d/|a|)^{1/2} = 2(d^3\epsilon_0 m/q^2)^{1/2}$, so $P = 8(d^3\epsilon_0 m/q^2)^{1/2}$.

☐ ELECTROSTATIC POTENTIAL AND CAPACITANCE

S241. The potential at A is

$$V_A = \frac{1}{4\pi\epsilon_0}\left(\frac{q_1}{3d/4} + \frac{q_2}{3d/4}\right) = 0$$

since $q_2 = -q_1$. At B we have

$$V_B = \frac{1}{4\pi\epsilon_0}\left(\frac{q_1}{2d/3} + \frac{q_2}{d/3}\right) = \frac{3}{4\pi\epsilon_0}\frac{1}{d}\left(\frac{q_1}{2} + q_2\right).$$

Substituting we find $V_B = 3 \times 9 \times 10^9 \times (1/12) \times (2.5 \times 10^{-8} - 5 \times 10^{-8}) = -56.3$ volts.

S242. For uniform fields we have $\Delta V = Ed$. With $E = 2 \times 10^4$ N/C and $d = 2$ cm we find $\Delta V = 2 \times 10^4 \times 0.02 = 400$ volts.

S243. If the Earth is electrically neutral, we have only kinetic (T) and gravitational potential energy (U). Conservation of energy thus gives $T + U = $ constant. At infinity both T and U are zero, so the constant here is zero, and hence

$$0 = T - \frac{2GM_em_p}{R_e + h}.$$

Thus

$$\begin{aligned}
T &= 2GM_em_p/(R_e + h) \\
&= 2 \times 6.7 \times 10^{-11} \times 6 \times 10^{24} \times 1.67 \times 10^{-27}/(6.5 \times 10^6) \\
&= 2 \times 10^{-19} \text{ J}.
\end{aligned}$$

If the Earth is positively charged the particle must do work against the electrical potential $V(r) = Q_e/4\pi\epsilon_0 r$, so conservation of energy now requires $T + U + qV = $ constant. At infinity we have $T = 0, U = V = 0$, and the particle will just fail to reach the Earth's surface if $T = 0$ at $r = R_e$. Thus the minimum charge Q_e on the Earth is given by

$$T = \frac{Q_e e}{4\pi\epsilon_0 R_e} - \frac{2GM_em_p}{R_e}.$$

Using the expression for T found above we note that the second term on the rhs is $T(R_e + h)/R_e$. This gives $Q_e = 4\pi\epsilon_0 T(2R_e + h)/e = 1.8 \times 10^{-3}$ C.

S244. The closest approach is achieved when the particle is incident head-on: conservation of energy (cf. the previous answer) gives

$$\frac{1}{2}mv^2 = \frac{Ze^2}{4\pi\epsilon_0 b}$$

where b is the closest approach distance. Thus

$$b = \frac{Ze^2}{2\pi\epsilon_0 mv^2}.$$

The stationary particle behaves as if it had "size" b and cross-sectional area $\sigma \sim \pi b^2$. In an electrically charged gas (a plasma) v can be related to the temperature, and σ can be used to estimate properties such as thermal conductivity, etc.

S245. When the particles are at rest their total linear momentum and energy are both zero (no kinetic energy and negligible potential energy). Momentum is conserved as there are no external forces, so that

$$m_1 v_1 + m_2 v_2 = 0$$

where v_1 and v_2 are the velocity components of the two particles when they are a distance L apart. No work is done on the system, so the sum of the kinetic and electrostatic potential energies is conserved at zero, i.e.

$$\frac{1}{2}m_1 v_1^2 + \frac{1}{2}m_2 v_2^2 + \frac{q_1 q_2}{4\pi\epsilon_0 L} = 0.$$

Substituting v_2 from the first equation into the second we get:

$$(1 + \frac{m_1}{m_2})m_1 v_1^2 = \frac{-q_1 q_2}{2\pi\epsilon_0 L},$$

and using the data given

$$v_1^2 = \frac{e^2}{20\pi\epsilon_0 L m_p}.$$

Thus, choosing $v_1 > 0$ we have $v_1 = 5.25 \times 10^3\,\mathrm{m\,s^{-1}}$, $v_2 = -(m_1/m_2)v_1 = -2.1 \times 10^4\,\mathrm{m\,s^{-1}}$. The particles' relative velocity is $v_1 - v_2 = 2.63 \times 10^4\,\mathrm{m\,s^{-1}}$.

S246. From the definition, $1\,\mathrm{eV} = 1.6 \times 10^{-19} \times 1 = 1.6 \times 10^{-19}\,\mathrm{J}$, we can find the required potential difference ΔV from

$$\Delta V = \frac{E}{q},$$

where E is the energy. Measuring E in eV and q in electron charges gives ΔV in volts. Thus $\Delta V = 10^5/2 = 5 \times 10^4$ volts.

S247. The potential at P is

$$V(\mathrm{P}) = \Sigma_i \frac{q_i}{4\pi\epsilon_0 d_i},$$

where d_i is the distance of the charge q_i from P. Since $d_i = [(x_i - 2)^2 + (y_i - 2)^2]^{1/2}$, we find

$$V(\mathrm{P}) = \frac{10^{-6}}{4\pi\epsilon_0}\left(\frac{1}{[2^2 + 2^2]^{1/2}} + \frac{2}{[1^2 + 2^2]^{1/2}} - \frac{3}{[1^2 + 2^2]^{1/2}}\right) = -843\,\mathrm{V}.$$

S248. Since the field is uniform we have $\Delta V = Ed = E_0 y_1$. With $E_0 = 100\,\mathrm{N/C}$ and $y_1 = 5\,\mathrm{cm} = 0.05\,\mathrm{m}$, we find $\Delta V = 100 \times 0.05 = 5$ volts.

We can calculate the work W using the formula $W = Fd\cos\theta$, where F is the constant electrostatic force, d the straight-line distance moved, and θ the angle between the path and the force. In the present case we must exert a force $F = E_0 Q$ to drag the charge quasistatically in the negative y-direction, and the work done in the two cases is

$$W_1 = E_0 Q_0 y_1,$$

and

$$W_2 = E_0 Q_0 (x_1^2 + y_1^2)^{1/2} \cos\theta,$$

using $d = y_1$ in the first case and $d = (x_1^2 + y_1^2)^{1/2}$ in the second. Substituting $\cos\theta = y_1(x_1^2 + y_1^2)^{-1/2}$, we see that $W_2 = E_0 Q_0 y_1$, so that in both cases the work done is $W_1 = W_2 = E_0 y_1 Q_0 = 5$ J. The same result follows immediately from the energy conservation law $W = U_f - U_i$, where U_f, U_i are the final and initial potential energies, as $U_f - U_i = Q_0 \Delta V$.

S249. For a point charge we have

$$E(r) = \frac{1}{4\pi\epsilon_0} \frac{Q_0}{r^2},$$

$$V(r) = \frac{1}{4\pi\epsilon_0} \frac{Q_0}{r}.$$

Dividing these two equations gives $V/E = r$, so $r = 500/100 = 5$ m. Using this value in the formula for V gives $Q_0 = 4\pi\epsilon_0 r V = 2.78 \times 10^{-7}$ C.

S250. The potential difference $\Delta V = V_B - V_A$ between A and B is just minus the field multiplied by the distance AB, i.e. $\Delta V = -Ed = -2000$ V.

(a) The charge q is negative, so work must be done to move it to lower potential. The total work done is $W = (\text{force}) \times (\text{distance moved}) = |qEd| = |q\Delta V| = 20$ J.

(b) The work done moving a charge in a static electric field depends only on the endpoints of the path, and not on its shape, so the charge from A to B by any other route, including the one specified here, is exactly the same as along AB, i.e. 20 J.

S251. The potential of the charged shell is $V = \mathcal{E} = 10^3$ V. Since $V = Q/(4\pi\epsilon_0 a)$ we have $Q = 4\pi\epsilon_0 V a = 1.1 \times 10^{-6}$ C. The work done is $W = qV = 10^{-6} \times 10^3 = 10^{-3}$ J. If the charge penetrates the shell, no extra work is required to bring it to the center, as the potential is constant inside the shell.

S252. For a charged spherical shell we have $V = q/(4\pi\epsilon_0 r)$ so $q = 4\pi\epsilon_0 V r$. By conservation of charge $Q = 1000q = 4000\pi\epsilon_0 V r$. The total volume of the merged drop must be the sum of the individual volumes, as mercury is incompressible, so $4\pi R^3/3 = 1000 \times 4\pi r^3/3$, i.e. $R = 10r$. Thus

$$V_1 = \frac{Q}{4\pi\epsilon_0 R} = \frac{4000\pi\epsilon_0 V r}{40\pi\epsilon_0} = 100V.$$

The electrostatic energy of a spherical conductor is

$$U = \frac{q^2}{8\pi\epsilon_0 r}.$$

Thus for the separated drops $U = 1000(q^2/8\pi\epsilon_0 r) = 2000\pi\epsilon_0 V^2 r$. The merged drop has $U_1 = Q^2/(8\pi\epsilon_0 R) = 2 \times 10^6 \pi\epsilon_0 V^2 r$. This is 1000 times the original energy U, because work had to be performed against the electrostatic repulsion to merge the drops.

S253. The minimum approach distance is given by a head-on trajectory on which the alpha particle is brought to a halt. Thus all of the kinetic energy E_α is transformed into potential energy, so

$$E_\alpha = \frac{1}{4\pi\epsilon_0} \frac{q_\alpha q_{Au}}{d}$$

giving $d = 79e^2/(\pi\epsilon_0 E_\alpha)$. Now $E_\alpha = 1$ MeV $= 10^6$ eV $= 1.6 \times 10^{-13}$ J. Substituting this we get $d = 4.6 \times 10^{-13}$ m. At $x = 2d$ the potential energy is

$$U = \frac{1}{4\pi\epsilon_0} \frac{q_\alpha q_{Au}}{2d} = \frac{E_\alpha}{2}.$$

The kinetic energy there is thus $T_\alpha = E_\alpha - U = E_\alpha/2 = U$, so the ratio is 1.

S254. From the definition the kinetic energy of the electron is 1 keV $= 1.6 \times 10^{-16}$ J. Setting this equal to $m_e v^2/2$ we find $v = (2E_e/m_e)^{1/2} = 1.9 \times 10^7$ m s^{-1} (or 68 million km/h!). The momentum of such an electron is $m_e v = 1.7 \times 10^{-23}$ Ns. The momentum of each electron is destroyed on hitting the electrode, so a flux of $n = 10^{10}$ s^{-1} on the electrode produces a force of $nm_e v = 1.7 \times 10^{-13}$ N only. If the electrons are replaced by protons the velocity of each goes down by a factor $(m_p/m_e)^{1/2} \approx 43$, so the speed becomes $v_p = 4.4 \times 10^5$ m s^{-1}. The momentum $m_p v_p$ goes up by the same factor, giving a total force of 7.3×10^{-12} N.

S255. A current of $I = 10^{-4}$ A flowing for 1 s deposits charge $Q = 10^{-4}$ C. The total number of electrons is $N = Q/e = 6.25 \times 10^{14}$. The total energy deposited is $E = NE_e$, where $E_e = 10^{10}$ eV is the energy of each electron. Using 1 eV $= 1.6 \times 10^{-19}$ J, we find $E = 10^6$ J.

S256. Using $Q = CV$, we substitute for C and $V = \mathcal{E}$ to find $Q = 10^{-8} \times 10^3 = 10^{-5}$ C. The energy is

$$U = \frac{CV^2}{2} = 10^{-8} \times 10^6/2 = 5 \times 10^{-3} \text{ J}.$$

The capacitance of a parallel plate capacitor is $C = \epsilon_0 A/d$, where A is the area of the plates and d the distance between them. Thus doubling the distance halves the capacitance, which becomes $C' = C/2 = 5 \times 10^{-9}$ F. The total charge Q cannot change, so the new potential difference is $V' = Q/C' = 2V$. Thus the new energy is

$$U' = \frac{C'V'^2}{2} = \frac{(C/2)(2V)^2}{2} = CV^2 = 2U = 10^{-2} \text{ J}.$$

The extra energy came from the work done separating the plates.

S257. After connection, the potentials V_1, V_s of the electrometer and sphere are equal, i.e. $V_1 = V_s = 900$ V. Thus by definition we have

$$Q = CV_0 \tag{1}$$

$$Q_1 = CV_1 = CV_s \tag{2}$$

The charge Q_s on the sphere is related to its potential V_s by

$$V_s = \frac{Q_s}{4\pi\epsilon_0 r}. \tag{3}$$

By charge conservation we have

$$Q = Q_1 + Q_s, \tag{4}$$

or, substituting from (1–3),

$$CV_0 = CV_s + 4\pi\epsilon_0 r V_s.$$

Since $V_s = V_1$, we have immediately that

$$C = 4\pi\epsilon_0 r \frac{V_1}{V_0 - V_1} = \frac{1}{9 \times 10^9} \times 3 \times 10^{-2} \frac{900}{1350 - 900} = 6.67 \times 10^{-12} \text{ F}.$$

From (1) and (2) we get $Q = CV_0 = 9 \times 10^{-9}$ C and $Q_1 = CV_1 = 6 \times 10^{-9}$ C.

S258. The equivalent capacitance is

$$C_T = \frac{1}{1/C_1 + 1/C_2} = \frac{C_1 C_2}{C_1 + C_2} = \frac{8 \times 24}{8 + 24} = 6 \ \mu\text{F}.$$

The charge on the equivalent capacitor is thus

$$Q_T = C_T \mathcal{E} = 6 \times 60 \ \mu\text{C} = 3.6 \times 10^{-4} \text{ C}.$$

The capacitors are in series, so this is also the charge on each of them, i.e. $Q_1 = Q_2 = Q_T$. The individual capacitances then give the potential differences

$$V_1 = \frac{Q_1}{C_1} = 45 \text{ V},$$

$$V_2 = \frac{Q_2}{C_2} = 15 \text{ V}.$$

(Note that these values satisfy $V_1 + V_2 = \mathcal{E}$, as they must.) The stored electrostatic energy is

$$U_T = \frac{Q_T^2}{2C_T} = 0.011 \text{ J}.$$

After the dielectric is removed, the capacitance C_2 is decreased by the factor K_d so that its new value C_2' is $C_2' = C_2/K_d = 8 \ \mu\text{F}$. The two capacitors are now equal, making the calculation easier. Thus $V_1 = V_2 = \mathcal{E}/2 = 30$ V, and $U_T = 2U_1 = 2(C_1 V_1^2/2) = 7.2 \times 10^{-3}$ J.

S259. The total capacitance of the two capacitors connected in parallel after the circuit is closed is $C_T = C_1 + C_2 = C_1 + 2C_1 = 3C_1$. The total charge is conserved, i.e. $Q_T = Q$, so the voltage on both capacitors will be

$$V_1 = V_2 = \frac{Q_T}{C_T} = \frac{Q}{3C_1}.$$

The charges on each are then

$$Q_1 = C_1 V_1 = \frac{Q}{3},$$

$$Q_2 = C_2 V_2 = 2C_1 V_1 = \frac{2Q}{3},$$

and the energies are

$$U_1 = \frac{C_1 V_1^2}{2} = \frac{Q^2}{18C_1},$$

$$U_2 = \frac{C_2 V_2^2}{2} = C_1 V_1^2 = \frac{Q^2}{9C_1}.$$

Thus $U_T = Q^2/6C_1$. Initially we had $U_T = U_1 = Q^2/2C_1$, which was larger. Energy was released in sharing the charge out between the two capacitors (currents dissipate heat).

S260. If the level of dielectric liquid has fallen a distance $vt = h < l$ we have two capacitors in parallel, i.e.

$$C_1(t) = \epsilon_0 \frac{lh}{l/100} = 100\epsilon_0 vt,$$

$$C_2(t) = K_d \epsilon_0 \frac{(l-h)l}{l/100} = 200\epsilon_0(l - vt).$$

Thus $C(t) = C_1 + C_2 = 100\epsilon_0(2l - vt)$ until $t = l/v$, when $C(t)$ stays constant at $C = 100\epsilon_0 l$. The charge is just $C(t)V$.

S261. A parallel plate capacitor has capacitance $C = \epsilon A/x$, where $\epsilon = K_d\epsilon_0$, so the energy stored is

$$U = \frac{CV^2}{2} = \frac{Q^2}{2C} = \frac{Q^2 x}{2\epsilon A}.$$

This increases as the plates are separated, so work must be done. Since work = force × (distance moved in direction of force), the force between the plates must be

$$F = \frac{U}{x} = \frac{Q^2}{2\epsilon A}.$$

(We are justified in dividing U by x to get F, since F turns out to be constant.) With the data given, $F = 1.89 \times 10^{12} Q^2$ N. If the capacitor is not to collapse, this must be less than the weight of 200 kg, i.e. $200 \times 9.8 = 1960$ N. Thus we require $Q < 3.2 \times 10^{-5}$ C. The force scales as

$$F \propto \frac{Q^2}{\epsilon} \propto \frac{Q^2}{K_d},$$

so if K_d is halved the maximum possible charge goes down by a factor $\sqrt{2}$, to 2.3×10^{-5} C.

S262. (a) We have $Q = CV, C = \dfrac{\epsilon_0 S}{d}$ and $E = \dfrac{Q}{\epsilon_0 S}$,

so

$$Q = \frac{\epsilon_0 SV}{d}, \quad E = \frac{V}{d}$$

and

$$U = \frac{CV^2}{2} = \frac{\epsilon_0 SV^2}{d}.$$

(b) The capacitance now changes to $C = \epsilon_0 S/2d$, while the voltage remains unchanged, so

$$Q = \frac{\epsilon_0 SV}{2d}, \quad E = \frac{V}{2d}$$

and

$$U = \frac{\epsilon_0 SV^2}{2d}.$$

(c) This case is identical to (b)!

S263. After the spheres are connected, charge will flow until the spheres are at the same potential. The potential of a conducting sphere with charge Q and radius R is $V = Q/4\pi\epsilon_0 R$, so charge flows until the charges on the spheres are Q_1, Q_2, with $Q_1/R_1 = Q_2/R_2$ or

$$Q_1 = Q_2 R_1/R_2. \tag{1}$$

Moreover charge must be conserved in the flow, so that

$$Q_1 + Q_2 = q_1 + q_2.$$

Hence eliminating Q_1 we find

$$Q_2\left(1 + \frac{R_1}{R_2}\right) = q_1 + q_2,$$

or

$$Q_2 = \frac{(q_1 + q_2)R_2}{R_1 + R_2}.$$

Then (1) gives

$$Q_1 = \frac{(q_1 + q_2)R_1}{R_1 + R_2}.$$

With the values given we get $Q_1 = 2.67 \times 10^{-8}$ C, $Q_2 = 1.33 \times 10^{-8}$ C.

S264. Each conducting sphere is a capacitor, so that the stored electrical energy is $U = CV^2/2 = QV/2$, where C, V, Q are the capacitance, potential and charge. Since $V = Q/4\pi\epsilon_0 R$ for a sphere of radius R, we have total energy

$$U_i = \frac{1}{8\pi\epsilon_0}\left(\frac{q_1^2}{R_1} + \frac{q_2^2}{R_2}\right)$$

before the spheres are connected, and

$$U_f = \frac{1}{8\pi\epsilon_0}\left(\frac{Q_1^2}{R_1} + \frac{Q_2^2}{R_2}\right)$$

after connection. Substituting the data from the previous problem and its answer we find $U_i = 9.9 \times 10^{-5}$ J, $U_f = 2.4 \times 10^{-5}$ J. As can be seen, the final energy is lower. This is to be expected, as the currents flowing in the connected system must dissipate some energy as heat.

S265. The first sphere will accumulate charge q_1 such that its potential $V_1 = q_1/4\pi\epsilon_0 R_1$ reaches the external potential V. Thus $q_1 = 4\pi\epsilon_0 V R_1 = 10^{-6}$ C. When the two spheres are connected, charge will flow until the two potentials are equal, i.e. they will have charges Q_1, Q_2 with $Q_1/R_1 = Q_2/R_2$ and $Q_1 + Q_2 = q_1$. Thus $Q_2 = 2Q_1$ and $Q_1 + Q_2 = 10^{-6}$ C, implying $Q_1 = 3.33 \times 10^{-7}$ C, $Q_2 = 6.66 \times 10^{-7}$ C.

S266. (a) Here the shells are independent, so

$$V_1 = \frac{q}{4\pi\epsilon_0 a},$$

$$V_2 = \frac{q}{12\pi\epsilon_0 a},$$

and the potential difference is

$$\Delta V = V_1 - V_2 = \frac{q}{6\pi\epsilon_0 a}.$$

(b) See the Figure. The potential of the inner sphere has the value $V_1' = q/(4\pi\epsilon_0 a)$ resulting from its own charge, plus the potential V_2 of the outer sphere. Hence $V_1 = V_1' + V_2$, so

$$\Delta V = V_1 - V_2 = V_1' = \frac{q}{4\pi\epsilon_0 a}.$$

(The outer sphere behaves as if it had a total charge $2q$, so that its potential is

$$V_2 = (1/4\pi\epsilon_0)(2q/3a).)$$

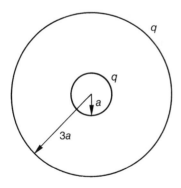

S267. The field inside a perfect conductor must vanish, so by Gauss's law, charges $-q$ and $+q$ are induced on the inner and outer surfaces of the shell respectively. Thus

$$E(r_{\text{out}}) = \frac{q}{4\pi\epsilon_0 r_{\text{out}}^2}, \quad E(r_c) = 0$$

and

$$E(r_{in}) = \frac{q}{4\pi\epsilon_0 r_{in}^2}.$$

The potentials follow by superposition, i.e.

$$V(r_{out}) = \frac{q}{4\pi\epsilon_0 r_{out}},$$

$$V(r_c) = \frac{1}{4\pi\epsilon_0}\left(\frac{q}{r_c} - \frac{q}{r_c} + \frac{q}{2R}\right) = \frac{q}{8\pi\epsilon_0 R},$$

and

$$V(r_{in}) = \frac{1}{4\pi\epsilon_0}\left(\frac{q}{r_{in}} - \frac{q}{R} + \frac{q}{2R}\right) = \frac{q}{4\pi\epsilon_0}\left(\frac{q}{r_{in}} - \frac{q}{2R}\right).$$

If the shell is grounded its potential is zero, so that the charge on its outer surface vanishes. However, Gauss's law still requires a charge $-q$ on the inner surface. The fields and potentials are calculated as above, but now with no charge on the shell's outer surface. Thus

$$E(r_{out}) = E(r_c) = 0, \quad E(r_{in}) = \frac{q}{4\pi\epsilon_0 r_{in}^2},$$

and

$$V(r_{out}) = V(r_c) = 0,$$

$$V(r_{in}) = \frac{1}{4\pi\epsilon_0}\left(\frac{q}{r_{in}} - \frac{q}{R}\right).$$

S268. We can regard the capacitor as the superposition of two parallel capacitors at the same voltage, with one containing the dielectric. Their capacitances are

$$C_1 = \frac{K_d\epsilon_0 a}{d}\left(\frac{a}{2} - x\right)$$

and

$$C_2 = \frac{\epsilon_0 a}{d}\left(\frac{a}{2} + x\right).$$

Thus

$$C(x) = C_1 + C_2 = \frac{\epsilon_0}{d}\left[\frac{a^2}{2}(K_d + 1) + ax(1 - K_d)\right].$$

With $K_d = 2$ this gives $C(x) = (\epsilon_0/d)(3a^2/2 - ax)$, which reduces to $C = \epsilon_0 a^2/d$ for $x = a/2$ (all the dielectric removed) as it should.

To find the current we need the charge

$$Q(x) = C(x)V = \frac{\epsilon_0 V}{d}\left(\frac{3}{2}a^2 - ax\right).$$

In a time interval Δt the dielectric moves a distance $\Delta x = u\Delta t$. The charge changes by $\Delta Q = -\epsilon_0 Va\Delta x/d$ (i.e. it decreases). Thus

$$I = \frac{\Delta Q}{\Delta t} = -\frac{\epsilon_0 Va}{d}\frac{\Delta x}{\Delta t} = -\frac{\epsilon_0 Vau}{d}$$

S269. The large distance between C and the AB system allows us to assume that they do not influence each other. Then

$$V_C = \frac{1}{4\pi\epsilon_0}\frac{-2q}{R} = -\frac{q}{2\pi\epsilon_0 R},$$

$$V_B = \frac{1}{4\pi\epsilon_0}\frac{q}{2R} + \frac{1}{4\pi\epsilon_0}\frac{q}{2R} = \frac{q}{4\pi\epsilon_0 R},$$

$$V_A = \frac{1}{4\pi\epsilon_0}\frac{q}{R} + \frac{1}{4\pi\epsilon_0}\frac{q}{2R} = \frac{3q}{8\pi\epsilon_0 R},$$

where we have used the fact that the potential is constant inside a spherical shell in writing the last equation. After B and C are connected, charge flows between them until their potentials become equal. If the new charges are Q_B, Q_C, conservation of charge gives

$$Q_B + Q_C = q - 2q = -q. \tag{1}$$

Since the new V_B, V_C are equal,

$$\frac{1}{4\pi\epsilon_0}\frac{q + Q_B}{2R} = \frac{1}{4\pi\epsilon_0}\frac{Q_C}{R},$$

or

$$q + Q_B = 2Q_C. \tag{2}$$

(1,2) are two equations for Q_B, Q_C, with the solution $Q_C = 0, Q_B = -q$. The potentials become

$$V_A = \frac{1}{4\pi\epsilon_0}\frac{q}{R} - \frac{1}{4\pi\epsilon_0}\frac{q}{2R} = \frac{q}{8\pi\epsilon_0 R},$$

$$V_B = \frac{1}{4\pi\epsilon_0}\frac{q - q}{2R} = 0,$$

$$V_C = 0.$$

S270. The cow's hooves are at lower (ground) potential than the fence, so current flows. Birds sitting on the wire have both feet at the same potential, so no current flows through them. In principle one could grasp the fence provided one's feet were off the ground; but you should *not* attempt this as it is potentially fatal.

☐ ELECTRIC CURRENTS AND CIRCUITS

S271. The current I is charge/time, i.e. 0.5 A. This is equivalent to $Q/e = 3.13 \times 10^{18}$ electrons per second. The power supplied by the battery is $P = I\mathcal{E} = 6$ W.

S272. The voltage drop V_0 across the battery is given by the emf minus the voltage required to drive the current I through the internal resistance, i.e.

$$V_0 = \mathcal{E} - R_{\text{in}}I.$$

Thus $R_{\text{in}} = (\mathcal{E} - V_0)/I = (6 - 5.8)/0.2 = 1\Omega$.

S273. (a) From the Figures, for resistors in series (Case 1), we have that the total voltage drop $I(r + 2R)$ must equal $\mathcal{E}$, so that

$$I = \frac{\mathcal{E}}{r + 2R} = 2 \text{ A}.$$

(b) If the resistors are connected in parallel, their total resistance is $(2/R)^{-1} = R/2$, so that

$$I = \frac{\mathcal{E}}{r + R/2} = 5 \text{ A}.$$

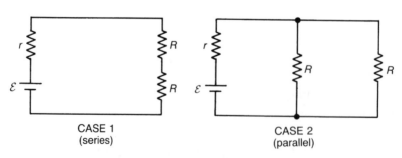

CASE 1
(series)

CASE 2
(parallel)

S274. By definition, the resistance is

$$R = \rho\frac{l}{A} = \rho\frac{l}{\pi(r_2^2 - r_1^2)} = 1.75 \times 10^{-6}\frac{10}{\pi(1 - 0.81) \times 10^{-4}} = 0.29 \ \Omega.$$

S275. The resistance is given by $R = \rho l / A$.
(a) We get $R_1 = 1.75 \times 10^{-6} \times 0.1/(3 \times 10^{-6}) = 5.8 \times 10^{-2}$ Ω.
(b) Here $R_2 = \rho l / \pi r^2 = 5.6 \times 10^{-4}$ Ω.

S276. Choosing the current directions as in the Figure, Kirchhoff's laws give

$$I_3 = I_2 - I_1 \tag{1}$$

$$R_1 I_1 + R_2 I_2 - \mathcal{E} = 0 \tag{2}$$

$$R_x I_3 + R_2 I_2 + \mathcal{E} = 0. \tag{3}$$

Here and in subsequent problems the curving arrows indicate the direction in which we apply Kirchhoff's 2nd law.

Using (1) to eliminate I_3 and substituting $R_1 = R_2 = R$, we get

$$I_1 + I_2 = \frac{\mathcal{E}}{R}$$

$$\left(1 + \frac{R_x}{R}\right) I_2 - \frac{R_x}{R} I_1 = -\frac{\mathcal{E}}{R}.$$

Eliminating I_1 we find

$$I_2 = \frac{\mathcal{E}}{R} \frac{R_x - R}{R + 2R_x} = 3 \frac{R_x - 2}{2R_x + 2} \text{ A.}$$

Clearly the current I_2 vanishes if $R_x = 2\Omega$.

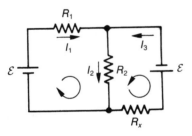

S277. Using Kirchhoff's laws with the current directions shown in the Figure gives

$$I_1 = I_2 + I_3, \tag{1}$$

$$\mathcal{E}_1 = R_3 I_3 + R_1 I_1$$

$$\mathcal{E}_2 = R_2 I_2 - R_3 I_3.$$

Substituting the values given, the last two equations become

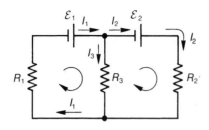

$$7 = 8I_3 + 4I_1$$

$$3 = 5I_2 - 8I_3.$$

Using (1) we find

$$7 = 12I_3 + 4I_2, \qquad (2)$$

$$3 = 5I_2 - 8I_3. \qquad (3)$$

Multiplying (2) by 2 and (3) by 3 and adding gives $I_2 = 1$ A, so from (2) or (3) $I_3 = 0.25$ A, and from (1) $I_1 = 1.25$ A.

S278. Using Kirchhoff's laws

$$i_3 = i_1 + i_2$$

$$\mathcal{E}_1 = i_2 R_1 + i_3 R_3$$

$$-\mathcal{E}_2 = i_1 R_2 + i_3 R_3.$$

With the values of $R_1, R_2, \mathcal{E}_1, \mathcal{E}_2$ given, the first equation simplifies the other two to

$$4i_1 + 7i_2 = 3$$

$$3i_1 + 2i_2 = -1,$$

which have the solution $i_1 = -1$ A, $i_2 = 1$ A. There is *no* current in the resistor R_3, as $i_3 = i_1 + i_2 = 0$.

S279. The current in the original circuit is $I = (2\mathcal{E} - \mathcal{E})/4R = \mathcal{E}/4R$ clockwise. Thus the voltage drop between A and B is $V_A - V_B = -IR + 2\mathcal{E} = 7\mathcal{E}/4$. The emf X must be in the same direction as the two in the original circuit, with magnitude $X = 7\mathcal{E}/4$.

S280. For the case shown in Figure 1, we have $V_{ab} = V_1$. But by Kirchhoff's laws

$$V_{ab} = I_1(R_A + R),$$

so

$$V_1 = I_1(R_A + R),$$

or

$$R = \frac{V_1}{I_1} - R_A.$$

This case, therefore, gives R close to V_1/I_1 if $R_A \ll R$.

For the case shown in Figure 2, $V_{ab} = V_2$, and Kirchhoff's laws give

$$\frac{V_{ab}}{R} + \frac{V_{ab}}{R_V} = I_2.$$

Thus

$$\frac{V_2}{R} = I_2 - \frac{V_2}{R_V}.$$

Solving this equation for R gives

$$R = \frac{V_2}{I_2 - V_2/R_V}.$$

This case gives R close to V_2/I_2, if we can neglect the second term in the denominator, i.e. if $V_2/R_V \ll I_2$. This is equivalent to requiring $R_V \gg V_2/I_2 \approx R$, so here we need $R_V \gg R$.

Thus the two methods will both give R close to (measured voltage)/(measured current), if $R_V \gg R \gg R_A$.

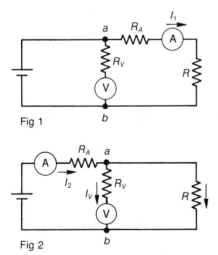

Fig 1

Fig 2

S281. With the voltmeter connected across cb we have the circuit shown in the Figure. Let I be the current in the main circuit (i.e. flowing from the power supply), and let I_1 be the current between the points c and b. The voltmeter will measure $V_{cb} = RI_1$. To find I_1 we first calculate the equivalent resistance of the whole circuit:

$$R_T = R + \frac{1}{1/R + 1/r} = \frac{2r + R}{r + R}R.$$

Thus

$$I = \frac{\mathcal{E}}{R_T} = \frac{\mathcal{E}(r + R)}{R(2r + R)}. \tag{1}$$

We can find I_1 from the fact that the potential drop through the resistor between c and b must be the same as that through the voltmeter between the same points, i.e.

$$RI_1 = r(I - I_1).$$

Solving for I_1 we get

$$I_1 = \frac{r}{R + r}I.$$

Substituting for I from (1) we get

$$I_1 = \frac{r}{r + R} \times \frac{\mathcal{E}(r + R)}{R(2r + R)} = \mathcal{E}\frac{r}{(2r + R)}$$

and therefore

$$V_{cb} = RI_1 = \mathcal{E}\frac{r}{2r + R}.$$

By symmetry we get the same result for V_{ba}.

Note that if $r \gg R$ we have $V_{cb} = V_{ba} \approx \mathcal{E}/2$, very close to the value in the circuit without the voltmeter. However, if the internal resistance r is not much larger than the resistances R, the voltmeter will draw a significant current and thus reduce the voltage drop V_{cb} or V_{ba} below this value.

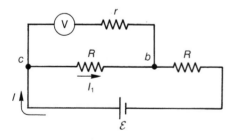

S282. The equivalent resistance in each of the three cases is

$$R_a = \frac{1}{1/R + 1/R} = \frac{R}{2},$$

$$R_b = R + R = 2R,$$

and

$$R_c = R + \frac{1}{1/R + 1/R} = \frac{3R}{2}.$$

The dissipated power is

$$P = \frac{\mathcal{E}^2}{R}.$$

Thus

$$P_a = 2\frac{\mathcal{E}^2}{R}, \quad P_b = \frac{1}{2}\frac{\mathcal{E}^2}{R}, \quad P_c = \frac{2}{3}\frac{\mathcal{E}^2}{R}.$$

The dissipated power is largest in circuit (a).

S283. The power dissipated is $P = I^2 R = V^2/R = 242$ W. The total energy used is $E = Pt = 8.7 \times 10^5$ J $= 0.242$ kWh.

S284. The total energy used is $E = Pt = 0.1 \times 24 = 2.4$ kWh. The cost is therefore $2.4 \times 30 = 72$ cents.

S285. The power is $P = IV = 3.6$ kW. The total energy is $E = Pt = 432$ kJ or 0.12 kWh.

S286. (a) When the switches are open, we have a single circuit with a current

$$I = \frac{\mathcal{E}}{R_1 + R_2 + R_3} = 1.5 \text{ A}.$$

(b) When both switches are closed, the resistor R_2 is shorted out, so the equivalent circuit is as shown in the Figure. The current through the ammeter is again $I = 1.5$ A, so the potential difference between a and b is

$$V_{ab} = IR_3 = 4.5 \text{ V}.$$

But since a and b are also connected through the power supply and resistor R_1, we also have

$$V_{ab} = \mathcal{E} - I_1 R_1.$$

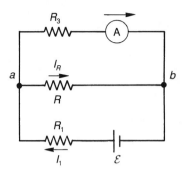

Thus $I_1 = (\mathcal{E} - V_{ab})/R_1 = 2.5$ A. The current in R is $I_R = I_1 - I = 1$ A (the net current at a must vanish), so

$$R = \frac{V_{ab}}{I_R} = 4.5 \ \Omega.$$

S287. Let each bulb have resistance R.

(a) If the bulbs are connected in series the total resistance is $8R$, the current $I_S = \mathcal{E}/8R$ and the dissipated power is $P_S = \mathcal{E}I_S = \mathcal{E}^2/8R$.

(b) If the bulbs are connected in parallel the total resistance R_P is given by $R_P^{-1} = 8/R$, so that $R_P = R/8$ and the current is $I_P = 8\mathcal{E}/R$. The dissipated power is $P_P = \mathcal{E}I_P = 8\mathcal{E}^2/R$.

The son is right, by a factor 64.

S288. The circuit is shown in the Figure. With the telephone connected there are two resistors in parallel, so the resistance is

$$\frac{1}{R_c} = \frac{1}{R_s + 2xr} + \frac{1}{R_T + 2(d - x)r} \tag{1}$$

With the telephone disconnected we have simply

$$R_d = 2xr + R_s \tag{2}$$

We recognize this in the first term in (1), so

$$\frac{1}{R_c} = \frac{1}{R_d} + \frac{1}{R_T + 2(d - x)r}.$$

Thus

$$R_T + 2(d - x)r = \frac{R_c R_d}{R_d - R_c},$$

which can be solved for x to give

$$x = \frac{R_T}{2r} + d - \frac{R_c R_d}{2r(R_d - R_c)}.$$

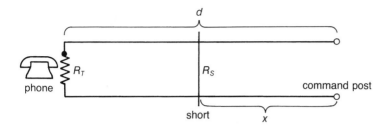

With the data given we find $x = 3.0\,$km. From (2) we find $R_s = R_d - 2xr = 114\,\Omega$.

S289. If the bulbs are connected in parallel the total resistance R_T is given by

$$\frac{1}{R_T} = \frac{1}{R} + \frac{1}{2R} = \frac{3}{2R},$$

so the total current is $I = \mathcal{E}/R_T = 3\mathcal{E}/2R$ where $\mathcal{E}$ is the mains voltage. The currents I_A, I_B through the bulbs obey

$$\frac{I_A}{I_B} = \frac{2R}{R} = 2$$

and Kirchhoff's laws require

$$I_A + I_B = I = \frac{3\mathcal{E}}{2R},$$

so $I_A = \mathcal{E}/R$ and $I_B = \mathcal{E}/2R$. The emitted powers are $P_A = \mathcal{E}I_A = \mathcal{E}^2/R$, $P_B = \mathcal{E}I_B = \mathcal{E}^2/2R$ and the total power is $P = P_A + P_B = 3\mathcal{E}^2/2R$. If the bulbs are connected in series, the total resistance is $R_T = R + 2R = 3R$, and the current is $I = \mathcal{E}/3R$. The powers are $P_A = I_A^2 R = \mathcal{E}^2/9R$, $P_B = I_B^2.2R = 2\mathcal{E}^2/9R$, and the total power is $P = \mathcal{E}^2/3R$. Thus bulb A is brighter when the connection is in parallel, which also maximizes the total power output. The two clerks can agree.

S290. In the first case no current flows in the circuit involving $\mathcal{E}_2$; the current in the circuit involving $\mathcal{E}_1$ is

$$I_1 = \frac{\mathcal{E}_1}{R_{AB}} = 0.1 \text{ A}.$$

The resistance of the interval AP is

$$R_{AP} = R_{AB}(AP/AB) = 20 \times (60/100) = 12 \ \Omega,$$

so $V_{AP} = I_1 R_{AP} = 1.2$ V. This must equal the potential difference $\mathcal{E}_2$ given by the power supply, i.e. $\mathcal{E}_2 = 1.2$ V. Also $V_R = 0$, since there is no current in R.

In the second case, the connection P is exactly in the middle of the resistor AB, thus splitting it into two $r = 10\,\Omega$ resistors, and producing the equivalent circuits shown in the Figure. Using Kirchhoff's laws we have

$$I_3 = I_1 + I_2 \tag{1}$$

$$\mathcal{E}_2 = I_3 r + I_2 R \tag{2}$$

$$\mathcal{E}_1 = I_3 r + I_1 r. \tag{3}$$

Using the known values of $\mathcal{E}_1, \mathcal{E}_2, R$ and r, and putting (1) in (2) and (3), we get

$$1.2 = 10(I_1 + I_2) + 30I_2 = 10I_1 + 40I_2$$

$$2 = 10(I_1 + I_2) + 10I_1 = 20I_1 + 10I_2$$

with the solution $I_2 = 0.4/70 = 5.7 \times 10^{-3}\,\text{A} = 5.7\,\text{mA}$. This is the milli-ammeter reading. Similarly $I_1 = 97.1\,\text{mA}$. Finally, $V_R = I_2 R = 0.171\,\text{V}$.

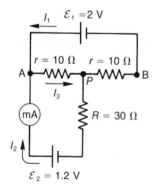

S291. Once they are fully charged the capacitors draw no current. Thus in a steady state

$$I = \frac{\mathcal{E}}{r + R_1 + R_2} = 4\,\text{A}.$$

The potential difference across the capacitors is thus $V = I(R_1 + R_2) = 10.8\,\text{V}$. As the capacitors are connected in series the effective capacitance C is given by

$$\frac{1}{C} = \frac{1}{C_1} + \frac{1}{C_2},$$

so that $C = 0.0143\,\mu\text{F}$. This gives $Q = CV = 0.154\,\mu\text{C}$, and $Q_1 = Q_2 = Q$. When the switch is closed the steady-state current remains the same, $I = 4\,\text{A}$.

But the capacitors are no longer connected in series, so we have $V_1 = IR_1 = 6$ V and $V_2 = IR_2 = 4.8$ V. Thus $Q_1 = C_1 V_1 = 0.3$ μC and $Q_2 = C_2 V_2 = 0.096$ μC.

S292. The current flowing in both resistors is the same:

$$I_1 = I_2 = I = \frac{\mathcal{E}}{R_1 + R_2} = 2 \text{ A}.$$

The potential difference V_{AB} follows from the voltage drop across R_2: $V_{AB} = IR_2 = 8$ V. Thus $Q_2 = C_2 V_2 = C_2 V_{AB} = 40\,\mu$C. Also $Q_1 = C_1 V_1 = C_1 V_{AB} = 8$ μC.

☐ MAGNETIC FORCES AND FIELDS

S293. We take the origin of coordinates half-way between the two wires, the x-axis perpendicular to them and the y-axis parallel to them (see Figure). Each wire produces a magnetic field acting in circles centered on it and thus in the $\pm z$-direction at points in the x, y plane. With the orientations shown in the Figure both fields point *into* the plane ($-z$-direction) between the two wires, so the total field is the sum:

$$B(x) = \frac{\mu_0}{2\pi} \frac{2I}{d + 2x} + \frac{\mu_0}{2\pi} \frac{2I}{d - 2x},$$

i.e.

$$B(x) = \frac{\mu_0}{2\pi} \frac{4Id}{d^2 - 4x^2},$$

for $-d/2 < x < d/2$. With the data given $B(x) = 8 \times 10^{-7}(1 - 4x^2)^{-1}$ T. At $x = 0$, $B = 8 \times 10^{-7}$ T and the force is $F = evB = ecB/2 = 1.92 \times 10^{-17}$ N, acting in the x-direction if the velocity is in the y-direction. If the velocity is reversed the force points in the $-x$-direction.

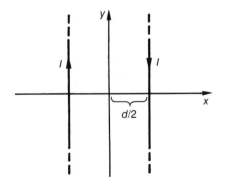

S294. The magnetic force between the wires is $F_m = \mu_0 I_1 I_2/(2\pi d)$ per unit length, and the weight per unit length is $W = mg$. In equilibrium (see Figure) the tension T in a cable must satisfy $T\cos\theta = W$, $T\sin\theta = F_m$, so

$$\tan\theta = \frac{F_m}{mg} = \frac{\mu_0 I_1 I_2}{2\pi mgd}. \tag{1}$$

We can eliminate d since $\sin\theta = d/(2a)$. Using the fact that $d \ll a$ we see that θ is also small, so that $\sin\theta \approx \tan\theta$. Hence substituting $d \approx 2a\tan\theta$ into (1) gives $\tan^2\theta = \mu_0 I_1 I_2/4\pi ga$. With the data given we find $\tan\theta = (2 \times 10^{-7}/9.8)^{1/2} = 1.43 \times 10^{-4}$, so that $\theta = 8.2 \times 10^{-3\circ}$. The magnetic field at the midpoint is the superposition of the fields produced by each wire, i.e.

$$B = \frac{\mu_0}{2\pi}\left(\frac{I_1}{d/2} + \frac{I_2}{d/2}\right),$$

where both fields point vertically downwards. Using $d/2 = a\tan\theta$ we find $B = 2 \times 10^{-7} \times 3/(1.43 \times 10^{-4}) = 4.2 \times 10^{-3}$ T.

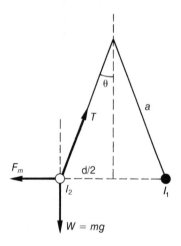

S295. By Ampère's law the field vanishes outside the coil. By symmetry it is circular (clockwise, by the right hand rule) inside the coil, and its magnitude depends only on r. Using Ampère's law for a circular path inside the coil (see Figure) gives

$$\frac{1}{\mu_0}B(r)2\pi r = NI, \quad a < r < b.$$

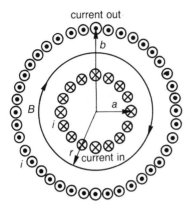

i.e.

$$B(r) = \frac{\mu_0}{2\pi} \frac{NI}{r}.$$

With the data given we find $B(r = 1.5 \text{ m}) = 2 \times 10^{-7} \times 10^4/1.5 = 1.33 \times 10^{-3}$ T.

S296. We can safely use the formula for an infinite solenoid, i.e. $B = \mu_0 n I$, where n is the number of turns per unit length, since the solenoid is slender (i.e. much longer than its radius). Since the currents in the two windings are $I_1 = I = 2$ A, $I_2 = -I = -2$ A, we find the total field

$$B = B_1 + B_2 = \frac{\mu_0}{l}(N_1 - N_2)I = -2.51 \times 10^{-3} \text{ T}.$$

The minus sign shows that the field acts in the direction given by applying the right hand screw rule to the *second* layer.

S297. The arm will strike the bell if the magnetic torque $\Gamma = -\mu B \sin \theta$ exceeds the critical value Γ_s, where B is the field produced by the solenoid. For a solenoid of n turns per unit length, we have $B_s = \mu_0 n I$, where I is the current. But by Ohm's law, $I = \mathcal{E}/nlr$, so since $B = 0.01 B_s$,

$$\Gamma = -0.01 \mu \mu_0 \frac{\mathcal{E}}{rl} \sin \theta, \tag{1}$$

independent of n. With the data given, $T = 1.07 \times 10^{-6}$ N.m. This is less than the critical value $\Gamma_s = 10^{-5}$ N.m, so the bell will not work. We see from (1) that a higher voltage battery or stronger permanent magnet is required.

S298. The field B_P is the superposition of the fields of the two loops. The formula for the field on the axis of a single loop is

$$B = \frac{\mu_0 I r^2}{2(r^2 + x^2)^{3/2}},$$

where r is the loop radius and x the distance along the axis from the center of the loop (the sign is determined by the right hand rule; see Figure). Using this with the data given yields $B_P = B_1 - B_2 =$

$$\frac{\mu_0 I (2r_0)^2}{2(4r_0^2 + 4r_0^2)^{3/2}} - \frac{\mu_0 I r_0^2}{2(r_0^2 + 4r_0^2)^{3/2}} = \left(\frac{4}{2 \times 8^{3/2}} - \frac{1}{2 \times 5^{3/2}} \right) \frac{\mu_0 I}{r_0} = 0.044 \frac{\mu_0 I}{r_0}.$$

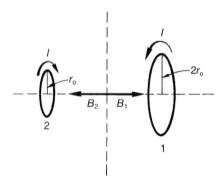

S299. The magnetic field of the long wire points everywhere into the plane of the loop (see Figure), with magnitude

$$B(x) = \frac{\mu_0}{2\pi} \frac{I}{x}, \tag{1}$$

where x is measured from the wire to the loop. By symmetry the forces on sides AB and CD of the loop cancel out, and the forces F_{AC}, F_{BD} on AC, BD only have x-components. With the current directions shown (see Figure) we find the resultant force

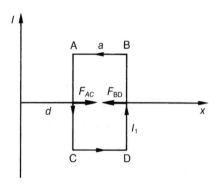

$$F = F_{AC} + F_{BD} = I_1 B(d)b - I_1 B(d+a)b.$$

Using (1) we find

$$F = \frac{\mu_0 I I_1}{2\pi} b \left[\frac{1}{d} - \frac{1}{d+a} \right] \tag{2}$$

leading to a force $F = 1.067 \times 10^{-6}$ N directed away from the wire.

S300. If there are N turns on a coil, the rhs of equation (2) of the previous problem is multiplied by N. Each coil must supply a force $F = wb$ to balance the weight of the train, so from the modified equation (2) above we require

$$w = \frac{\mu_0 N I I_1}{2\pi} \left[\frac{1}{d} - \frac{1}{d+a} \right].$$

Since w is fixed, to minimize I and I_1 we need to maximize the term in square brackets. We can make the negative part of this term negligible by choosing $a \gg d$. Then the requirement simplifies to

$$w \approx \frac{\mu_0 N I I_1}{2\pi d}. \tag{1}$$

Evidently we will minimize I, I_1 by making N as large as possible and d as small as possible. (The latter requirement makes it very easy to arrange that $a \gg d$.) For the data given, (1) shows that $N = 5 \times 10^6 wd / I I_1 = 5000$.

S301. From equation (1) of the previous question, the condition for balance is $d \propto 1/w$. The football players increase w from $1000\,\mathrm{kg\,m^{-1}}$ to $1300\,\mathrm{kg\,m^{-1}}$, so d decreases from 1 cm to $1 \times 1000/1300 = 0.77$ cm.

S302. The magnetic field at a distance r from a very long straight wire carrying current I has circular symmetry about the wire and strength

$$B = \frac{\mu_0}{2\pi} \frac{I}{r}.$$

By symmetry it is clear that one half of the wire contributes exactly one half of this expression. The field at O is the superposition of two such half-infinite wires (at right angles), giving total field $B_s = \mu_0 I/(2\pi r)$, together with the field of a quarter-circle loop at its center. Since the field of a full circular loop at the center is $B = \mu_0 I/(2r)$, the quarter loop adds a contribution $B_l = \mu_0 I/(8r)$. Hence the total field at O is

$$B = B_s + B_l = \left(\frac{1}{2\pi} + \frac{1}{8} \right) \frac{\mu_0 I}{r} = 0.28 \times \frac{1.26 \times 10^{-6} \times 1}{0.1} = 3.53 \times 10^{-6} \text{ T}.$$

The direction of the field is fixed by the right hand rule (into the page).

S303. The magnetic force between the rod and the wire is

$$F_m = L\frac{\mu_0}{2\pi}\frac{I^2}{L} = \frac{\mu_0}{2\pi}I^2,$$

so the equilibrium condition $\Sigma F = -mg + F_m = 0$ becomes

$$\frac{\mu_0}{2\pi}I^2 - mg = 0,$$

so that $I = (2\pi mg/\mu_0)^{1/2}$.

If the current in the wire is doubled, F_m becomes $F'_m = \mu_0 I^2/\pi$, so Newton's second law $\Sigma F = F'_m - mg = ma$ gives $ma = 2mg - mg = mg$, i.e the initial acceleration a is exactly g, upwards.

S304. The force on the particle is qvB, directed perpendicular to the motion. This force can do no work, so the particle must move at constant speed in a circle, the magnetic force supplying the required centripetal force. If the radius of the circle is R, we must have

$$\frac{mv^2}{R} = qvB. \tag{1}$$

The angular frequency is defined as $\omega = v/R$, so from (1) we find directly that $\omega = qB/m$. This is called the *gyrofrequency, Larmor frequency* or *cyclotron frequency* of the particle. Charged particles gyrate about magnetic fieldlines at this characteristic frequency: note that it is independent of their velocity.

If the velocity is not in the plane perpendicular to the field, we can consider the instantaneous components $v_\perp, v_\parallel$ perpendicular and parallel to it. The parallel component $v_\parallel$ produces zero magnetic force, while $v_\perp$ as before produces a force perpendicular to the field and always directed towards a particular fieldline. Since there is no force component along the fieldline, the particle moves with constant velocity $v_\parallel$ along it while gyrating about it as before. The combination of these two motions is a spiral centered on the fieldline.

S305. The angular frequency ω (measured in rad s^{-1} is related to the circular frequency ν (measured in cycles/s = Hertz) by $\omega = 2\pi\nu$. The wavelength λ is given by this frequency as $\lambda = c/\nu$ with c the speed of light (see Chapter 3). Here $\omega = eB/m_e$ (see previous problem), so $\nu = eB/2\pi m_e$ and hence $\lambda = 2\pi m_e c/eB$. With the data given, $\lambda = 26\,\text{m}$.

S306. Taking P as the origin of the coordinate system shown in the Figure, at P we have

$$B_x = B_1 \sin\theta = \frac{\mu_0}{2\pi} \frac{I_1}{(a^2 + 4a^2)^{1/2}} \frac{a}{(a^2 + 4a^2)^{1/2}},$$

$$B_y = B_2 - B_1 \cos\theta = \frac{\mu_0}{2\pi} \frac{I_2}{2a} - \frac{\mu_0}{2\pi} \frac{I_1}{(a^2 + 4a^2)^{1/2}} \frac{2a}{(a^2 + 4a^2)^{1/2}},$$

$$B_z = \frac{\mu_0}{2\pi} \frac{I_3}{a},$$

Thus

$$B_x = \frac{\mu_0}{2\pi} \frac{I_1}{5a},$$

$$B_y = \frac{\mu_0}{2\pi} \frac{I_1}{10a},$$

$$B_z = \frac{\mu_0}{2\pi} \frac{I_1}{2a}.$$

Substituting the numerical values given we get $B_x = 3.2 \times 10^{-7}$ T, $B_y = 1.6 \times 10^{-7}$ T, $B_z = 8 \times 10^{-7}$ T, so $B = (B_x^2 + B_y^2 + B_z^2)^{1/2} = 8.76 \times 10^{-7}$ T.

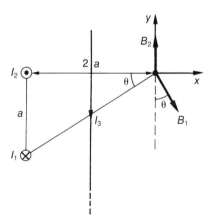

S307. The electric field exerts a constant force qE_0 in the direction on motion of the particle, and so performs total work qE_0d on it. This must all go into kinetic energy, so that the particle encounters the magnetic field region with velocity v given by

$$\tfrac{1}{2}mv^2 = qE_0d,$$

i.e. $v = (2qE_0d/m)^{1/2}$. The magnetic force acts perpendicular to the particle's motion and thus does no work on it, so that its speed remains constant and it moves in a circle (see e.g. S304). The radius R of the circle is fixed by the condition that the magnetic force qvB_0 should provide the centripetal force mv^2/R. Thus $qvB = mv^2/R$ or

$$R = \frac{mv}{qB_0} = \left(\frac{2mE_0d}{qB_0^2}\right)^{1/2}. \tag{1}$$

With B_0 as shown the particle will move up the page in a semi-circle, re-entering the electric field region at a point $2R$ above its entry point. For this distance to be d we require $2R = d$, i.e. $d = 2(2mE_0d/qB_0^2)^{1/2}$, giving $B_0 = (8mE_0/qd)^{1/2}$.

S308. Using equation (1) of the last solution, we find $D = 2R = 2(2mE_0d/qB_0^2)^{1/2}$ or $q/m = 8E_0d/(B_0^2D^2)$. With the data given we find $q/m = 9.67 \times 10^8$ C/kg. For an electron the corresponding ratio is $-e/m_e = -1.76 \times 10^{11}$ C/kg, and for a proton we get a ratio $e/m_p = 9.58 \times 10^8$ C/kg. The particle is probably a proton, as the deflection D is similar to that expected (making due allowance for experimental error). Note that the electron deflection would have the opposite sign, i.e. be on the opposite side of the initial track.

S309. Let the particle masses be m_1, m_2, m_3. Their velocities v_1, v_2, v_3 on entering the magnetic field region are given by energy conservation, i.e.

$$\frac{1}{2}m_1v_1^2 = qV$$

so that $v_1 = (2qV/m_1)^{1/2}$, etc. As the magnetic force acts perpendicular to the motion it does no work, so the velocities remain at these values. Each particle moves in a circle (see S304 and subsequent problems). The radii, etc. of the orbits follow from the equations of motion, in which the Lorentz force qv_1B, etc. must supply the centripetal force $m_1v_1^2/R_1$, so that

$$R_1 = \frac{m_1v_1}{qB} = \left(\frac{2V}{qB}\right)^{1/2}m_1^{1/2} \text{ etc.}$$

Thus the masses are in the ratios $m_1 : m_2 : m_3 = R_1^2 : R_2^2 : R_3^2 = 1 : 4 : 9$.

S310. The particle will begin to move in a circle of radius $R = mv/qB$ (see previous problems). Obviously if $R < b$ the particle will not reach $x = b$, and the condition for this is $v < v_c = bqB/m$. If v is larger than this, the particle will reach $x = b$ and continue in a straight line. From the Figure showing

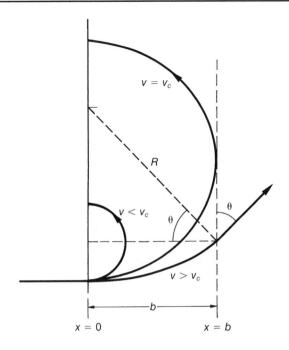

the trajectory for three values of v, we see that $\cos\theta = b/R = bqB/mv$. (Note that the first condition corresponds to $\cos\theta > 1$.)

S311. The total force on the particle is $q(E - vB)$, in the direction of E if the particle's charge $q > 0$ and in the opposite direction if $q < 0$. For the particle to be undeflected this force must vanish, so that $v = E/B$. If a narrow beam of charged particles with a range of velocities is directed into the apparatus, only those of precise speed v will remain undeflected and emerge from the far side. The apparatus can therefore be used as a velocity selector.

S312. As the solenoid is long and slender it is effectively infinite, so that the field inside it is $B = \mu_0 NI/L = 1.26 \times 10^{-2}$ T. Since AB, DC are parallel to the field direction, no forces act on them. The forces on AD, BC are equal and opposite and perpendicular to the loop plane. The forces are each $F = iBl$, where $l = AD = BC = 0.06$ m. Thus $F = 1 \times 1.26 \times 10^{-2} \times 0.06 = 7.56 \times 10^{-4}$ N. These forces form a couple, so the torque is $\Gamma = Fd$, where $d = AB = 0.1$ m. Thus $\Gamma = 7.56 \times 10^{-5}$ N m.

S313. The magnetic forces on the short sides of the loop act in opposite directions along the rotation axis and cancel each other. The long sides experience equal and opposite forces BIl perpendicular to the field, the directions being given by the right-hand screw rule. These form a torque $T = BIlw\cos\theta$ (see Figure). If I remains constant, this torque vanishes and then reverses at $\theta = 90°, 270°$, so the loop comes to rest there. But if I is reversed each time the loop passes through $\theta = 90°, 270°$ the torque acts in the same sense for all

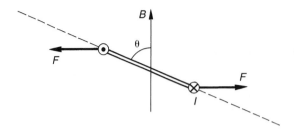

angles, so the loop continues to revolve. This is the principle of the DC electric motor.

S314. The forces acting on the mass are shown in the Figure. The magnetic Lorentz force quB, where u is the velocity, acts normal to the plane and the mass's motion (as the magnetic force always does). Assuming that q is small enough that the mass does not leave the plane, the acceleration in the plane is unaffected, and is given by Newton's second law as

$$a = \frac{F}{M} = g \sin \theta.$$

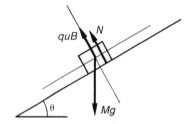

If the plane is not smooth the magnetic force will change a by changing N and thus the frictional force.

☐ ELECTROMAGNETIC INDUCTION

S315. Let x be the distance of the leading side of the loop from the boundary of the magnetic field region. Then the magnetic flux through the loop is

$$\Phi = B_0 l_1 (l_2 - x)$$

for $0 < x < l_2$. For $x < 0$ all of the loop is in the field region, so the flux has the constant value $\Phi = B_0 l_1 l_2$, and for $x > l_2$ the flux is zero. Hence the flux changes only for $0 < x < l_2$, and induces an emf

$$\mathcal{E} = -\frac{\Delta \Phi}{\Delta t} = -B_0 l_1 \frac{\Delta x}{\Delta t} = -B_0 l_1 v$$

in this case ($\mathcal{E} = 0$ for other values of x). The induced current is $\mathcal{E}/R = B_0 l_1 v/R = 0.5 \times 3/1.5 = 1$ A.

S316. When the loop plane is perpendicular to the magnetic field the flux is $\Phi = BA$. This decreases as the loop rotates, inducing an emf in one direction. Once the loop plane becomes parallel to the field, the flux begins to increase, inducing an emf in the opposite direction. In the first half of the cycle the flux decreases from BA to zero. If the rotation rate is $N\,\mathrm{s}^{-1}$, this takes a time $1/(2N)$ s, so that the average induced emf is $\mathcal{E}_{\mathrm{ave}} = BA/(1/2N) = 2NBA$.

This is the principle of the AC generator employed in power stations. This uses many loops, thereby increasing the emf proportionally.

S317. The current in the solenoid is $\mathcal{E}_1/R_1$, so it produces an instantaneous magnetic field $B = \mu_0 n \mathcal{E}_1/R_1$. The flux mAB in the coil reduces from its maximum value to zero over half the AC cycle time $1/N$ s (see previous solution) before reversing symmetrically, producing an average emf $\mathcal{E}_2 = 2NA\mu_0 nm\mathcal{E}_1/R_1$.

This is the principle of the transformer, which allows a voltage supply to be altered at will. The disadvantage of a transformer is that there is always considerable dissipation in the coils.

S318. The emf induced in the moving rod is $\mathcal{E}_{BA} = B_0 l v$ (higher potential at B than A by Lenz's law). With the data given, $\mathcal{E}_{BA} = 10$ V, so the currents are $I_{KM} = \mathcal{E}_{BA}/R_{KM} = 10$ A, $I_{LN} = \mathcal{E}_{BA}/R_{LN} = 5$ A. The total power dissipated in the circuit is thus

$$P = I_{KM}^2 R_{KM} + I_{LN}^2 R_{LN} = 100 \times 1 + 25 \times 2 = 150 \text{ W}.$$

This power must be supplied mechanically by pulling the rod AB with force F such that $P = Fv$, as there are no other energy losses from the system. Hence $Fv = 150$ W, so $F = 150/5 = 30$ N.

S319. At time t the moving wire has traveled a distance vt in the plane of the bent wire, so the length of moving wire between the contacts is

$$l = l_0 + 2vt \tan 30° = l_0 + \frac{2}{\sqrt{3}} vt.$$

As the wire moves it increases the flux enclosed by the triangle, and so induces an emf

$$\mathcal{E} = B_0 v l = B_0 v \left(l_0 + \frac{2}{\sqrt{3}} vt \right).$$

At time $t = 5$ s this and the data given imply

$$\mathcal{E} = 1 \times 2[0.5 + (2/\sqrt{3}) \times 2 \times 5] = 24.1 \text{ V}.$$

The resistance of the triangular loop at any time is $R = 3lr$, which increases with time in exactly the same way as $\mathcal{E} \propto l$. Hence the current in the triangle is

$$I = \frac{B_0 lv}{3lr} = \frac{B_0 v}{3r} = 6.67 \text{ A},$$

and is independent of time.

S320. With the bob at height x the magnetic flux is $\Phi = Bwx$, so the induced emf is $\mathcal{E} = -Bwv$, where v is the speed. If the bob reaches height h, the kinematic formula $v^2 = v_0^2 - 2gh$ shows that the initial speed is $v_0 = \sqrt{2gh}$. The largest induced emf is thus

$$\mathcal{E}_{\max} = -Bw\sqrt{2gh}.$$

B cannot exceed 10^{-4} T, and could be lower if the slide is not oriented exactly perpendicular to the local magnetic field. With the values given for w, h we find $|\mathcal{E}_{\max}| = 1.4 \times 10^{-4}$ V. The voltmeter must be able to measure voltages of this order of magnitude. (Note that the magnetic force is always negligible compared with gravity.)

S321. The magnetic flux is $\Phi = BS(t) = BS_0(1 - \alpha t)$, so the rate of change, and thus the induced emf, is $\mathcal{E} = -\Delta\Phi/\Delta t = S_0 B\alpha$. The current direction is determined by Lenz's law. The strength of the current is $I(t) = \mathcal{E}/R$, where $R = 2\pi r(t)\rho$. With $r(t) = [S(t)/\pi]^{1/2}$, we find

$$I(t) = \frac{\alpha B}{2\rho}\left[\frac{S_0}{\pi(1 - \alpha t)}\right]^{1/2}.$$

S322. A flux $\Phi = NAB$ is removed in $t = 10^{-3}$ s, so the induced emf is $\mathcal{E} = NAB/t = 1.2 \times 10^5$ V. This produces a current $I = \mathcal{E}/R = NAB/(Rt)$ and the dissipated power is $P = \mathcal{E}^2/R = (NAB)^2/(R^3 t^2) = 1.2 \times 10^{10}$ W. The total work done is $W = Pt = (NAB)^2/(R^3 t) = 1.2 \times 10^7$ J.

This shows the very large mechanical power required to remove conductors rapidly from magnetic field regions, and the dangers of rapidly decaying fields.

S323. We have $\Delta\Phi = (B_1 - B_2)A$, so the induced emf is $\mathcal{E} = \Delta\Phi/t = 1 \times 0.01/0.001 = 10$ V. The current is $I = \mathcal{E}/r = 1000$ A, the dissipated power is $P = \mathcal{E}I = 10^4$ W and the total heat produced is $Pt = 10$ J. While this is not large, it is extremely localized, and the very high current $I = 1000$ A is very dangerous. People working in regions of high magnetic field are strongly advised not to wear any conducting loops (e.g. bangles, rings).

S324. The flux through the loop was $\Phi = NBA$ and was reduced to zero in time t, so the induced emf is $\mathcal{E} = NBA/t$. The current is $I = \mathcal{E}/R = NBA/(Rt)$ and the total charge passed was $Q = It = NBA/R$, so

$$B = \frac{QR}{NA} = \frac{2 \times 10^{-6} \times 10}{20 \times 10^{-4}} = 10^{-2} \text{ T}$$

with the data given.

S325. The induced emf V is given by

$$V = L\frac{\Delta I}{\Delta t},$$

where $\Delta I = 10$ A is the change in the current in time Δt. With the data given, we find $V = 18 \times 10/0.25 = 720$ V.

S326. The relation

$$V = L\frac{\Delta I}{\Delta t}$$

shows that $L = V/(\Delta I/\Delta t)$. Here $L = 20/50 = 0.4$ H.

S327. Using

$$L = \frac{N\Phi}{I}$$

we find

$$L = \frac{100 \times 10^{-5}}{5} = 2 \times 10^{-4} \text{ H} = 0.2 \text{ mH}.$$

S328. At time t the normal to the loop plane makes an angle $\theta = \omega t$ to the magnetic field direction (see Figure), where we have chosen to measure t from the instant when the normal is parallel to the field. The magnetic flux through the loop is therefore

$$\Phi = NAB\cos\omega t.$$

The induced emf $\mathcal{E}$ is minus the rate of change of Φ with time t. To find this we consider the small change $\Delta\Phi$ in Φ which occurs when t increases to $t + \Delta t$. We have

$$\Phi + \Delta\Phi = NAB\cos\omega(t + \Delta t) = NAB(\cos\omega t \cos\omega\Delta t - \sin\omega t \sin\omega\Delta t),$$

using the identity $\cos(a + b) = \cos a \cos b - \sin a \sin b$. Now since Δt is small, we have

$$\cos\omega\Delta t \approx 1,$$

$$\sin\omega\Delta t \approx \omega\Delta t,$$

where $\omega\Delta t$ is measured in radians. Then the first term on the rhs above is just Φ itself, so we find that in time Δt, Φ changes by an amount

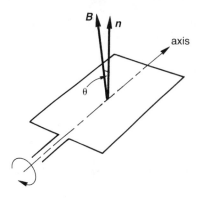

$$\Delta\Phi = -NAB\omega\Delta t \sin\omega t.$$

Thus

$$\mathcal{E} = -\frac{\Delta\Phi}{\Delta t} = NAB\omega \sin\omega t.$$

Hence the emf oscillates in time, and reverses once per revolution. It therefore produces an alternating current (AC). This is the principle of the *dynamo* or *generator*.

S329. The wind rotates the axle at angular velocity $\omega = v/(L/2) = 2v/L$. From the previous answer, the induced emf in the loop is $\mathcal{E}(t) = NBA\omega\sin\omega t = 2(NBA/L)v\sin\omega t$. The maximum voltage occurs when $\sin\omega t = 1$, and is

$$\mathcal{E}_{\max} = \frac{2NBAv}{L}.$$

With the data given, and using $v = 100$ km/h $= 27.8\,\mathrm{m\,s^{-1}}$, we find $\mathcal{E}_{\max} = 2 \times 200 \times 10^{-4} \times 0.1 \times 27.8/0.5 = 0.22$ V.

CHAPTER **THREE**

M ATTER AND WAVES

☐ **LIQUIDS AND GASES**

S330. In equilibrium the pressure of fluid in the left and right arms must be equal. By symmetry the water columns below the level of the oil are in balance, so we have to balance the oil column of height h against the remaining water column of height $(h - d)$ in the left arm (see Figure in the Problem), i.e.

$$\rho_0 h g = \rho_w (h - d) g,$$

where $\rho_w = 1000$ kg m^{-3} is the density of water. Thus

$$\rho_0 = \rho_w \frac{h - d}{h} = 0.8 \rho_w = 800 \text{ kg m}^{-3}.$$

When the second fluid is added, we must balance the oil column of height h against a column of the same height, but which is half water and half the second fluid (see Figure). Thus

$$\rho_0 g h = \rho_w g \frac{h}{2} + \rho_x g \frac{h}{2}.$$

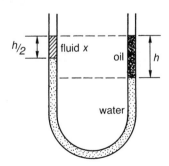

Multiplying each side by $2/h$ and rearranging we find

$$\rho_x = 2\rho_0 - \rho_w = (1600 - 1000) = 600 \text{ kg m}^{-3}.$$

S331. The hydrostatic pressure immediately below the large piston is $P_H = P_A + (M + M_l)g/A_l$, where P_A is the atmospheric pressure. In equilibrium this must equal the hydrostatic pressure P'_H a distance h below the small piston. Since $P'_H = P_A + mg/A_s + \rho_0 gh$, setting $P_H = P'_H$ gives

$$\frac{(M + M_l)g}{A_l} = \frac{mg}{A_s} + \rho_0 gh.$$

Rearranging, we find

$$m = (M + M_l)\frac{A_s}{A_l} - \rho_0 hA_s = 561 \times (10^{-4}/0.5) - 800 \times 1 \times 10^{-4} = 0.032 \text{ kg}.$$

S332. Any impurity will alter the density of the gold in the ring (usually lower it). The balance gives the ring's weight. Filling the volume measure to the brim and submerging the ring in it using the thread gives the ring volume when it is removed, so the density can be found. Archimedes is said to have been led to his principle by this type of experiment. (He was asked by the King of Syracuse to determine the purity of his crown: when he found it impure, the unfortunate goldsmith was executed.)

S333. (a) When standing, the woman's weight Mg is distributed over her shoe soles, of area roughly $2bl$. The pressure is thus $P \approx Mg/2bl \approx 16{,}800 \text{ N m}^{-2}$.

(b) When lying, the weight is distributed over an area $\approx hw$, so the pressure is $P \approx Mg/hw \approx 820 \text{ N m}^{-2}$. Lying on the floor is uncomfortable since much less of the body is in contact with it than in a bed, so the pressure is much higher on those areas.

The stiletto heels have area $A = 2 \times 10^{-4} \text{ m}^2$, so the pressure is $P = Mg/A \approx 3 \times 10^6 \text{ N m}^{-2}$. Even static pressures of this order are sufficient to cause damage to floors.

S334. The pressure gauge measures *excess* pressure, i.e. $P - P_A$, where P_A is atmospheric pressure, so it reads 6 atm. (It reads $P = 0$ before inflating, when the pressure inside the tire is clearly P_A!)

In equilibrium the road exerts a reaction force $P = 7P_A$ per unit area of tire in contact with it. This reaction pressure balances not only the weight per unit area of the rider and cycle, but also that of the atmosphere above. Exactly $1P_A$ is used for the latter purpose, so it is the excess pressure $6P_A$ which balances the weight. The tires deform so that a total area A is in contact with the road, and then

$$6P_A A = mg.$$

Thus $A = mg/6P_A = 70 \times 9.8/6 \times 10^5 = 1.1 \times 10^{-3}$ m^2, i.e. $A = 11$ cm^2.

S335. The cylinders are held together by the atmospheric pressure on their cross-sections. When M is maximal, the reaction force between the two cylinders vanishes, i.e. they are about to be pulled apart. Since the cables each have tensions $T = Mg$, horizontal equilibrium $\Sigma F_x = 0$ requires

$$AP_A = Mg,$$

so $M = AP_A/g = 1.02 \times 10^5$ kg $\simeq 102$ tonnes. For any other shape only the projected cross-sectional area is relevant (see S348). In a famous experiment teams of horses were unable to prise apart a pair of evacuated hemispheres ("the Magdeburg spheres").

S336. The buoyancy force F_B on the balloon and payload must balance their combined weight W. By Archimedes' principle $F_B = \rho_a V_b g$, and $W = (M_{\text{gas}} + m)g = \rho_b V_b g + mg$. Thus

$$\rho_b V_b g + mg = \rho_a V_b g,$$

or

$$\rho_b = \rho_a - \frac{m}{V_b} = \rho_a - 0.2 \text{ kg m}^{-3}.$$

Note that this is possible only if $\rho_a > \rho_{\text{crit}} = 0.2 \text{ kg m}^{-3}$, i.e. the balloon cannot be lifted to a height at which the air density is lower than the value $\rho_{\text{crit}} = m/V_b$. (This is effectively the average density of the balloon and payload.)

S337. By Archimedes' principle, the payload mass M plus the mass of supporting gas (H or He) must equal the mass of air displaced if the balloon is to rise, i.e.

$$M = V_{\text{H}}(\rho_a - \rho_{\text{H}}) = V_{\text{He}}(\rho_a - \rho_{\text{He}}),$$

when V_{H}, V_{He} are the required volumes of hydrogen and helium, so

$$\frac{V_{\text{He}}}{V_{\text{H}}} = \frac{\rho_a - \rho_{\text{H}}}{\rho_a - \rho_{\text{He}}} = \frac{1.3 - 0.09}{1.3 - 2 \times 0.09} = 1.08.$$

The volumes are not very different. The main reason for using hydrogen was the difficulty and expense of producing so much helium.

S338. Above the surface the ball falls under gravity, so using the kinematic formula $v^2 = v_0^2 - 2gy$ with $v_0 = 0$, we see that it enters the water ($y = -h$) with velocity $v = (2gh)^{1/2}$. When the ball is under the surface the resultant upward force acting on it is $F = \rho_w Vg - \rho_b Vg$, where V is its volume and $\rho_b = (2/3)\rho_w$ its density (i.e. buoyancy minus weight). Since its mass is $m = V\rho_b$ its upward acceleration is

$$a = \frac{F}{m} = \left(\frac{\rho_w}{\rho_b} - 1\right)g = 0.5g.$$

Using the kinematic formula $v^2 = v_0^2 + 2ay$ with initial velocity $v_0 = -(2gh)^{1/2}$, we find that the ball's downward motion is brought to a halt ($v = 0$) at a depth

$$y = -\frac{v_0^2}{2a} = -\frac{2gh}{2a} = -2h = -20 \text{ m}.$$

S339. (a) By Archimedes' principle, the buoy displaces its own weight of water whether inside or outside the yacht, so the water level remains unchanged.
(b) The anchor displaces its own weight of water when inside the boat, but less when it sinks (it just displaces its own *volume* of water, which weighs less). The water level drops.

S340. Let the cube be submerged to a depth x (see Figure). By Archimedes' principle the buoyancy force on the cube is $F_B = V_s \rho_w g$, where V_s is the submerged volume, i.e. $V_s = a^2 x$, and ρ_w is the density of water. In equilibrium F_B must balance the cube's weight $W = V\rho g$ with $V = a^3$. Thus from $F_B = W$ we find

$$a^2 x \rho_w = a^3 \rho,$$

i.e. $x = (\rho/\rho_w)a = 0.8 \times 0.05 = 0.04$ m. The submerged volume is therefore $V_s = a^2 x = (0.05)^2 \times 0.04 = 10^{-4}$ m^3. This is also the volume of the water displaced. The new height of the water is

$$h_{\text{new}} = \frac{V_0 + V_s}{A},$$

whereas the original height was

$$h_{\text{old}} = \frac{V_0}{A}.$$

Thus $h = h_{\text{new}} - h_{\text{old}} = V_s/A = 10^{-4}/10^{-2} = 10^{-2}$ m.
    When the mass m is added, the weight W is increased to $W' = W + mg = V\rho g + mg$. The buoyancy force becomes $F_B' = V\rho_w g$ as now the

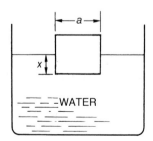

whole cube is submerged. Requiring $F'_B = W'$ for equilibrium as before, we find

$$V\rho_w g = V\rho g + mg,$$

so $m = (\rho_w - \rho)V = (1000 - 800) \times (0.05)^3 = 0.025$ kg $= 25$ g.

S341. By Archimedes' principle, the buoyancy force F_B on the cube when it is just submerged is (see Figure) $F_B = \rho_w Vg$, where ρ_w is the density of water and $V = a^3$ the volume of the submerged cube. This must balance the weight plus the downward force F, i.e.

$$\rho Vg + F = \rho_w Vg.$$

Thus

$$\rho = \rho_w - \frac{F}{a^3 g} = 1000 - \frac{3.43}{10^{-3} \times 9.8} = 650 \text{ kg m}^{-3}.$$

When the cube floats freely, it is submerged only to a depth h, say, so the submerged volume is $a^2 h$ and by Archimedes' principle the buoyancy force becomes $F_B = \rho_w a^2 hg$. This now balances just the weight $\rho a^3 g$ of the cube, so

$$\rho_w a^2 hg = \rho a^3 g,$$

or $h = (\rho/\rho_w)a = (650/1000) \times 0.1 = 0.065$ m.

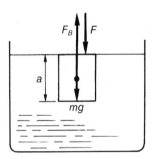

S342. The cube has total mass $M = 3a^3 \rho_w/4$, and will float when it displaces a mass M of water. Since $c \ll a$, the base area of the container is very close to a^2, so the water must reach a height $h = 3a/4$ (see Figure in the problem). The minimum volume of water needed to float the cube is thus

$$V \simeq 4 \times ah\frac{c}{2} = \frac{3}{2}a^2 c,$$

where we have considered only the water around the four sides of the cube.

We see that V becomes arbitrarily small as we decrease c: the pressure at a given depth below the surface of a continuous body of fluid is precisely the same no matter how little of it there is. In practice the lower limit on the volume of water occurs when there is so little of it that surface tension breaks it up and it is no longer a continuous fluid.

S343. By Archimedes' principle the buoyancy force is given by the combined weights of water and oil displaced (see Figure in problem), i.e.

$$F_B = \rho_w a^2 h g + \rho_o a^2 (a - h) g$$

The dynamometer reading W_D gives the force supplied by the spring, which must equal the difference between the cube's weight Mg and the buoyancy force F_B, i.e.

$$W_D = Mg - F_B.$$

Thus

$$M = \frac{W_D + F_B}{g} = \frac{W_D}{g} + a^2[\rho_w h + \rho_o(a - h)]$$

$$= 0.05 + (0.1)^2[1000 \times 0.02 + 500 \times 0.08] = 0.65 \text{ kg}.$$

The hydrostatic pressure P at the base of the cube is given by the depths of oil and water above that level, i.e. $P = \rho_o g d + \rho_w g h = 1176 \text{ N m}^{-2}$.

S344. (a) The iceberg's volume is $V = (h + x_s)^3$, so that its mass is $M = \rho_i V = \rho_i(h + x_s)^3$, and its weight is $W = Mg$. By Archimedes' principle this must equal the weight of seawater displaced, which is $M'g = \rho_s V'g$, where $V' = x_s(h + x_s)^2$ is the submerged volume. Equating M and M' we find

$$\rho_i(h + x_s)^3 = \rho_s x_s(h + x_s)^2$$

which gives

$$\rho_i(h + x_s) = \rho_s x_s$$

so that

$$x_s = \frac{\rho_i h}{\rho_s - \rho_i}$$

and thus with the data given $x_s = 5.625$ m.

(b) In fresh water the iceberg displaces a mass $\rho_f(h + x_s)^2 x_f$, which by Archimedes' principle again must equal $M = \rho_i(h + x_s)^3$. Thus $x_f = 0.9(2.5 + x_s)$ m, and using x_s from the answer to (a), we find $x_f = 7.313$ m. Since the side of the iceberg is $2.5 + x_s = 8.125$ m, only 81.25 cm or one-tenth is above the surface.

S345. The upward force exerted by surface tension (see Figure) is $F_t = 2\pi r \gamma \cos \theta$. This must balance the weight $W = \pi r^2 h \rho g$ of the column of liquid, i.e. $F_t = W$. Thus

$$h = \frac{2\gamma \cos \theta}{\rho g r}. \tag{1}$$

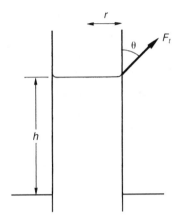

S346. Using equation (1) of the previous answer with $\theta = 0$, surface tension can hold a column of sap of height $h = 2 \times 0.07/(10^3 \times 9.8 \times 10^{-5}) = 1.4$ m. As trees grow considerably taller than this, capillary action cannot be significant.

S347. Assume that a very thin film of water fills the gap between the cap and the tube. Neglecting the mass of the water, the upward surface tension force $F_t = 2\pi r \gamma$ must balance the weight mg of the cap plus the reaction R of the tube (see Figure). Now $m = \pi r^2 d \rho$, so $F_t = mg + R$ implies

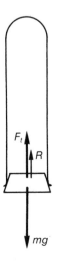

$$2\pi r\gamma = \pi r^2 d\rho g + R.$$

or

$$r = \frac{2\gamma}{d\rho g} - \frac{R}{\pi r d\rho g}.$$

Since R is positive, the largest r is given by $R = 0$, i.e. when the surface tension force just balances the weight of the cap. This gives $r_{max} = 2\gamma/(d\rho g) = 2 \times 0.07/(2 \times 10^{-3} \times 700 \times 9.8) = 0.01$ m $= 1$ cm.

S348. Imagine the sphere cut in half. The total outward pressure force on one hemisphere is given by the pressure difference $P_i - P_o$ multiplied by the *projected* area πr^2 of the hemisphere, because all components of this force other than the perpendicular outward one cancel by symmetry. This outward force must be balanced by the tension in the membrane, which by definition is $2\pi rt$. Thus $(P_i - P_o)\pi r^2 = 2\pi rt$, or

$$P_i - P_o = \frac{2t}{r}. \tag{1}$$

Since the liquid walls have both an inner and an outer surface, the total tension t is twice the surface tension, i.e. $t = 2\gamma$, and

$$P_i - P_o = \frac{4\gamma}{r}. \tag{2}$$

S349. Consider a length l of the tube, the excess pressure inside the section of tube being maintained by inserting bungs in either end. Imagine the tube now cut in half along its axis. The net outward pressure force again involves the projected area, and is thus $2rl(P_i - P_o)$. The tension force in the walls is $2lt$ (the bungs exert no tension), so equilibrium requires $2rl(P_i - P_o) = 2lt$ or

$$P_i - P_o = \frac{t}{r}. \tag{1}$$

As before, if we consider *surface* tension we have $t = 2\gamma$ as there are two surfaces, so

$$P_i - P_o = \frac{2\gamma}{r}. \tag{2}$$

In both cases we see that the tension required to contain a given pressure difference $P_i - P_o$ varies as the curvature radius r. *Along* the cylinder we have $r = \infty$, so boiling frankfurters split here first. This is why boiling frankfurters tend to split lengthways.

S350. A short section of the tire can be regarded as straight, so the considerations of the previous question apply. With $P_i = 7$ atm, $P_o = 1$ atm, we use equation (1) of the previous answer to get $t = (P_i - P_o)r = 6P_A r = 6 \times 10^5 \times 1.5 \times 10^{-2} = 9000 \, \text{N m}^{-1}$ with the data given.

S351. When the droplet is on the point of evaporating the surface tension force just balances the vapor pressure force. As the droplet has only an outer surface we use equation (1) of S348 with $t = \gamma$ to get $r = 2\gamma/(P_i - P_o) = 2\gamma/P_v = 2 \times 0.07/2300 = 6.1 \times 10^{-5} \text{m} = 6.1 \times 10^{-2}$ mm with the data given.

S352. The pressure P_i inside the balloon must obey

$$P_i - P_o = \frac{2t}{r}. \tag{1}$$

Initially $P_i = P_1, P_o = 8P_1/9$ and $r = r_1$, so

$$\frac{P_1}{9} = \frac{2t}{r_1}.$$

As P_o is reduced r increases. Its largest possible radius r_2 is given by setting $P_o = 0$ in (1). The pressure inside the balloon changes to P_2 because of expansion, so

$$P_2 = \frac{2t}{r_2}.$$

Dividing these two equations shows that

$$\frac{P_1}{P_2} = \frac{9r_2}{r_1}. \tag{1}$$

But since the temperature is fixed, $Pr^3 = \text{constant}$ (perfect gas law), i.e.

$$\frac{P_1}{P_2} = \frac{r_2^3}{r_1^3}. \tag{2}$$

Comparing these two equations we see that $r_2 = 3r_1$.

S353. The tension in the membrane must balance the pressure excess of the air sac, so from equation (1) of S348 we can write

$$(P_A - P_c)r = 2t.$$

In breathing out, both r and $P_A - P_c$ decrease (the latter because P_c increases and P_A is fixed). Equilibrium cannot be maintained unless t decreases. These changes are reversed in inhaling, so t increases. (If equilibrium failed in either state, the air sacs would either collapse or rupture.) The adjustment in t is provided by a protein – surfactant – which is very elastic. Asthma is associated with a failure of this mechanism to work properly.

S354. One might expect air to flow along the pipe to equalize the size of the two balloons. But amazingly, this is wrong: if the two balloons have spherical radii r_1, r_2 and interior air pressures P_{i1}, P_{i2} we must have

$$P_{i1} - P_o = \frac{2t_1}{r_1}, \tag{1}$$

$$P_{i2} - P_o = \frac{2t_2}{r_2}, \tag{2}$$

(cf. eqn (1) of S348). here P_o is the pressure in the enclosure and t_1, t_2 the surface tensions of the balloon material at radii r_1, r_2. These can be assumed constant ($t_1 = t_2 = t$) provided that each balloon is larger than the minimum radius r_{min}. Thus if $r_1 > r_2$ we must have $P_{i1} < P_{i2}$, i.e. the smaller balloon has a larger interior pressure (remember that it is hardest to blow up a balloon at the beginning, and this gets easier as the balloon expands!). Thus once the valve is opened, air will rush from the smaller balloon (making it smaller still) to the larger one (expanding it further). The air pressure inside the two connected balloons will equalize at some value P_i with $P_{i1} < P_i < P_{i2}$. Equations (1, 2) then require $t_2/t_1 = r_2/r_1 < 1$, i.e. the smaller balloon must contract below r_{min} and make $t_2 < t_1 = t$.

Note that even if we had started with two balloons with *equal* interior pressures P_{i1}, P_{i2}, a small perturbation making one of the pressures (say P_{i2}) even slightly larger than the other would have started this process off, and again we would have ended with one larger balloon (large r_1) and one small balloon with $r_2 < r_{min}$.

S355. Bernoulli's theorem states that the quantity

$$\frac{P}{\rho} + \frac{1}{2}v^2 + gh$$

is constant along a streamline in a fluid, where P, ρ, v are the fluid pressure, density and velocity and h the height of the point considered. Thus considering a streamline from the water surface (where v is effectively zero, $P = P_1$ and $h = H$) to the hole in the container (where the pressure is atmospheric, i.e. $P = P_A$), we have

$$\frac{P_1}{\rho} + gH = \frac{P_A}{\rho} + \frac{1}{2}v^2 + gh.$$

Thus the jet velocity v is given by

$$v^2 = 2g(H - h) + 2\frac{P_1 - P_A}{\rho}. \tag{1}$$

This gives $v = (2 \times 9.8 \times 1.5 + 2 \times 0.34 \times 10^5/10^3)^{1/2} = 9.87$ m s^{-1}.

If the container is open we can again use (1), but with $P_1 = P_A$. This gives $v = 5.42$ m s^{-1}.

S356. The blood pressure P varies with height y in the body as $P = P_0 + \rho g y$, where P_0 is the pressure at some reference height (e.g. the heart) and ρ is the blood density. To allow comparisons between measurements made on different occasions it is important to measure always at the same height y. Since it is often important to know the pressure in the heart it is convenient to choose the height of the heart as the standard. The upper arm is at a very similar height.

S357. Bernoulli's theorem states that

$$\frac{P}{\rho} + \frac{1}{2}v^2 + gy = \text{constant along streamlines.}$$

where P is the pressure, ρ the density, v the velocity and y the height of the water above some reference level. Taking this as the water surface we have $P = P_A$, the atmospheric value, and $v = y = 0$ there. At the other end of the pipe (in the gully), we have $y = -H$ (irrespective of the water height h, since H is much larger). Since water is incompressible, $\rho = $ constant, so the outflow velocity v from the pipe follows from

$$\frac{P_A}{\rho} = \frac{P_A}{\rho} + \frac{1}{2}v^2 - gH,$$

where we assume that the difference in atmospheric pressure between the two ends of the pipe is negligible. Thus $v = (2gH)^{1/2}$. This velocity remains constant, and carries off a mass $Q = \rho v a$ of water per unit time. The initial mass of water in the pool is $M = \rho h A$, so the time to drain the pool is

$$t = \frac{M}{Q} = \frac{Ah}{a(2gH)^{1/2}}.$$

With the data given, we find $t = 50 \times 2/[5 \times 10^{-4}(2 \times 9.8 \times 20)^{1/2}] = 10^4$ s $= 2.8$ h.

S358. If the pipe develops a leak at some point B above the water surface the water pressure will equal P_A there. Then with the same reference level as in the previous question, we will have

$$\frac{P_A}{\rho} = \frac{P_A}{\rho} + \frac{1}{2}v^2 + gy,$$

but now with $y > 0$. Thus $v^2 = -2gy < 0$, which can never be satisfied for any water velocity v. Hence there can be no streamline from the water surface to B, i.e. the flow stops. The effect of having air trapped in the pipe is to reduce its cross-sectional area a at some point, and hence reduce the flow rate. Siphons work well only if the pipe has no leaks and all of the air is carefully removed (e.g. here by filling the pipe from the lower end using a hosepipe before submerging the upper end in the pool).

S359. The water velocity follows from mass conservation: in one second the mass of water flowing with velocity v past a point where the pipe cross-section is A is $Q = \rho v A = \rho r$, where ρ is the water density. This must be constant in steady flow. Since ρ is constant (water is incompressible), this requires $r = vA =$ constant. Converting the water rate r to MKS units, $r = 6 \text{ m}^3 \text{min}^{-1} = 0.1 \text{ m}^3 \text{s}^{-1}$. Further, the values of A at the two ends of the pipe are $A_1 = \pi d_1^2 / 4 = 0.031 \text{ m}^2$ near the pump, and $A_2 = \pi d_2^2 / 4 = 0.126 \text{ m}^2$ at the other end. Thus $v_1 = r/A_1 = 3.2 \text{ m s}^{-1}$, and water leaves the pipe at velocity $v_2 = r/A_2 = 0.8 \text{ m s}^{-1}$.

S360. The pressure P_1 near the pump follows on using Bernoulli's theorem:

$$\frac{P_1}{\rho} + \frac{1}{2}v_1^2 = \frac{P_2}{\rho} + \frac{1}{2}v_2^2 + gh.$$

Since the upper end of the pipe is open to the atmosphere, $P_2 = P_A$, so

$$P_1 = \frac{1}{2}\rho(v_2^2 - v_1^2) + \rho gh + P_A$$

With the data given, the results of the previous problem, and $\rho = 10^3 \text{ kg m}^{-3}$, we find $P_1 = \frac{1}{2} \times 1000 \ (0.8^2 - 3.2^2) + 1000 \times 9.8 \times 20 + 10^5 = 2.91 \times 10^5 \text{ Nm}^{-2}$.

S361. Considering a streamline from the water surface down to the hole, Bernoulli's theorem gives

$$\frac{P_A}{\rho} + 0 + gh = \frac{P_A}{\rho} + \frac{1}{2}v^2 + 0,$$

where v is the (horizontal) velocity of the jet at the hole; this uses the facts that the pressure at both places is close to atmospheric, and the water velocity at the surface is very small because the container is wide. Thus $v = (2gh)^{1/2}$.

The jet is initially horizontal, but falls vertically from rest under gravity, so we can treat it like a projectile. Using $x = v_0 t + at^2/2$ with $v_0 = 0, a = -g$, the time to fall a distance $x = -(H - h)$ to the ground is

$$t = \left[\frac{2(H - h)}{g}\right]^{1/2}.$$

During this time the jet travels a horizontal distance

$$s = vt = 2[h(H - h)]^{1/2}.$$

Thus the jet has the biggest range ($\approx H$) when it is about halfway down the filled part of the container ($h \approx H/2$). Very short ranges result from holes near the water surface ($h \to 0$: little pressure head to drive the jet) and near the base of the container ($h \to H$: jet emerges too close to the ground).

S362. From Bernoulli's theorem we have

$$P + \frac{1}{2}\rho_w v^2 = P' + \frac{1}{2}\rho_w v'^2, \tag{1}$$

where P', v' are the pressure and velocity in the narrow section. Mass conservation, i.e.

$$\rho_w A v = \rho_w A' v'$$

gives $v' = v(A/A') = 4v$, so substituting this into (1) and rearranging we get

$$16v^2 - v^2 = \frac{2(P - P')}{\rho_w}$$

or $v^2 = (2/15)[(P - P')/\rho_w]$. But hydrostatic equilibrium of the mercury requires

$$P - P' = \rho_{Hg} g h,$$

so

$$v = \left(\frac{2}{15}\frac{\rho_{Hg}}{\rho_w} g h\right)^{1/2} = \left(\frac{2 \times 13{,}600 \times 9.8 \times 2.5 \times 10^{-2}}{15 \times 1000}\right)^{1/2} = 0.67 \text{ ms}^{-1}.$$

S363. Let the window and doorway have effective open cross-sectional areas A_w, A_d. If $A_d < A_w$, e.g. the door is only slightly ajar, any air draft entering the window must produce an air current with higher velocity on the open (outer) side of the door than the other side (see Figure). Hence by Bernoulli's theorem there is an excess pressure on the inside and the door slams. The

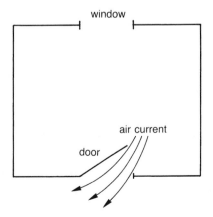

door will not slam if opened sufficiently wide as the pressure torque on the door is smaller than the frictional torque at the hinges.

S364. The upward force (*lift*) arises because the air must flow more swiftly over the airfoil than below it, lowering the air pressure there. The height difference between the top and bottom paths is negligible, so Bernoulli's theorem gives

$$\frac{P_1}{\rho} + \frac{v^2}{2} = \frac{P_2}{\rho} + \frac{m^2 v^2}{2},$$

where P_1, P_2 are the air pressures on the lower and upper surfaces. (Air is effectively incompressible if v is subsonic.) The speed above the airfoil is mv as the flow is steady. The pressure difference acting vertically upwards is thus

$$P_1 - P_2 = \frac{m^2 - 1}{2} \rho v^2.$$

The airplane will take off once the total lift force $F_L = A(P_1 - P_2)$ exceeds its weight Mg. Hence the minimum takeoff speed is given by

$$\frac{1}{2} A(m^2 - 1)\rho v^2 = Mg,$$

or

$$v^2 = \frac{2Mg}{(m^2 - 1)A\rho}. \tag{1}$$

With the data given, we find $v = [2 \times 500 \times 9.8/(0.21 \times 30)]^{1/2} = 39.4 \text{ m s}^{-1} = 142 \text{ km/h}$. At high-altitude airports, ρ is significantly smaller, and by (1) we see that the takeoff speed has to rise as $\rho^{-1/2}$.

The application of Bernoulli's theorem to airplane wings is subtle, as can be seen by considering the fact that airplanes can fly upside-down! The angle of the airplane to the horizontal (the *angle of attack*) is important in understanding this, as it determines the effective streamline ratio m.

S365. From equation (1) of the previous answer we have

$$\rho = \frac{2Mg}{(m^2 - 1)Av^2}.$$

Setting $v = v_{\max}$ gives the lowest density $\rho_{\min}$, which gives enough lift to support the airplane's weight. With the data given we get $\rho_{\min} = 2 \times 500 \times 9.8/(0.21 \times 30 \times 70^2) = 0.32 \text{ kg m}^{-1}$. Using the formula for $\rho(z)$ gives a maximum height $z_{\max} = -H \log_{10} \rho_{\min}$. Thus $z_{\max} = 23{,}000 \times 0.50 = 11{,}500 \text{ m} = 11.5 \text{ km}$.

S366. The main problem for early airplanes was the lack of sufficiently powerful engines to produce high takeoff speeds v. From equation (1) of S364 we see that v is reduced by making A large. This could have been achieved by increasing the wingspan, but it was difficult to produce a strong wing of great length. The easiest way of increasing A was to stack shorter wings above each other, i.e. biplanes (or triplanes).

S367. As in P364 the lift force is $F_L \propto Av^2$, where A is the wing area. At takeoff this just equals the weight Mg, where M is the bird's mass. Clearly A scales as l^2, while M scales as l^3, since the average densities of the species are the same. Thus $F_L \propto Av^2 \propto l^2v^2$, while $Mg \propto l^3$. Therefore $F_L = Mg$ requires $v \propto l^{1/2}$. Larger birds have higher takeoff speeds, and often have to run to achieve the necessary lift (e.g. flamingoes).

S368. This calculation here is essentially the same as in S364. The condition to lift the boat from the water is [cf. equation (1) of S364]

$$u^2 = \frac{2Mg}{(m^2 - 1)A_h\rho_w}.$$

The great difference here in comparison with P364 is that ρ_w is 1000 times larger than ρ for air. Thus even with u smaller than an airplane's takeoff speed, A_h can be made much smaller than A, i.e. hydrofoils are much smaller than airplane wings.

S369. Applied to the sail, Bernoulli's theorem gives

$$\frac{P_1}{\rho_a} + \frac{1}{2}u^2 = \frac{P_A}{\rho_a},$$

where P_1 is the pressure on the convex side of the sail, P_A is atmospheric pressure, and $u \approx w$ is the air speed along the convex side of the sail. This produces a force $F = (P_A - P_1)A \approx A\rho_aw^2/2$ acting towards the convex side (the air speed on the concave side is negligible), and so a force $F_1 = F \sin\theta \approx (A\rho_aw^2/2)\sin\theta$ in the direction of the yacht's motion.

If the wind comes from behind the boat, the sails are best deployed perpendicular to the wind velocity (see Figure). Since the yacht usually moves more slowly than the wind, essentially all of the wind's momentum is lost to the boat. Per unit area of the sails, the wind momentum is ρ_aw, and this arrives (and is lost) at velocity w. Hence the total wind momentum transferred to the boat per unit time is $\approx A\rho_aw^2$. By Newton's second law, this is the total force on the boat. The component F_2 of this in the forward direction is just given by multiplying by $\cos\phi$, giving $F_2 \approx A\rho w^2 \cos\phi$. For $\theta = \phi = 45°$, we have $\sin\theta = \cos\phi = 1/\sqrt{2}$ and

$$F_1 \approx \frac{A\rho_aw^2}{2\sqrt{2}}, \quad F_2 \approx \frac{A\rho_aw^2}{\sqrt{2}},$$

so that $F_1 \approx F_2/2$.

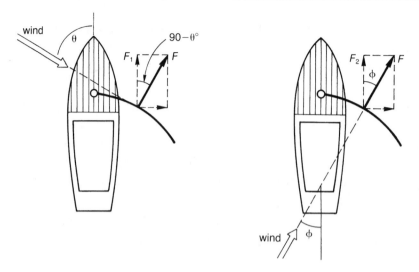

We see that these forces are comparable, so perhaps surprisingly, a yacht can sail at similar speeds ahead of and into the wind. Indeed, since the forward speed of the yacht increases the relative wind component w along the sail and thus F_1, suitably designed boats (e.g. sailboards) can actually sail *faster* into the wind than ahead of it: the speed in the latter case can never exceed the wind speed itself.

S370. The water mass flux encountering the front of the yacht is ρv per unit area, which has to be given speed $\approx v$ by the yacht, so the yacht has to supply momentum $F_d \approx A_f \rho v^2$ per unit time. By Newton's second law, F_d is the drag force on the boat. (In practice, the facts that the boat's wake is *turbulent* and the boat also excites waves makes the real drag force differ somewhat from the estimate F_d, but this is still adequate.) Setting $F_1 = F_d, F_2 = F_d$ in the two cases and using the expressions for F_1, F_2 from the previous solution shows that

$$v_1 \approx \left(\frac{A}{2A_f} \frac{\rho_a}{\rho} \sin \theta \right)^{1/2} w, \tag{1}$$

$$v_2 \approx \left(\frac{A}{A_f} \frac{\rho_a}{\rho} \cos \phi \right)^{1/2} w. \tag{2}$$

With the data given (1,2) yield

$$v_1 \approx \left(\frac{20}{0.6} \times 10^{-3} \times 0.71 \right)^{1/2} w = 0.15w,$$

$$v_2 \approx \left(\frac{20}{0.3} \times 10^{-3} \times 0.71 \right)^{1/2} w = 0.22w,$$

or $v_1 = 4.6$ km/h, $v_2 = 6.5$ km/h.

S371. By Archimedes' principle, the weight of the boat is equal to that of the water displaced, i.e. $Mg \approx A_f l \rho$. Also the sail area is $A = l^2/2$. Substituting for A_f, A into (1, 2) of the previous solution shows that for any angles θ, ϕ

$$v_1, v_2 \propto \left(\frac{l^3 \rho_a}{M}\right)^{1/2} w.$$

Since ρ_a, w are fixed, high sailing speeds are achieved by making l^3/M as large as possible. Long slender yachts are much faster than short stubby ones.

S372. The drag force resisting the sideways motion is $\approx A_s \rho s^2$, where s is the sideways velocity component. Equating this to the sideways forces $F_1 \cos \theta, F_2 \sin \phi$ (cf. S369) shows that s will be as small as possible if $A_s \rho \gg A \rho_a$. To give high forward speed, equations (1, 2) of S370 show that $A \rho_a$ should be as large as possible in comparison with $A_f \rho$. These two requirements are only compatible if $A_s \gg A_f$. Again we see that an efficient yacht should be slender. A_s is made large in practice by making the keel deep, as this also gives stability against the tendency of the wind pressure on the sails to push the boat over.

S373. The maximum speed v is fixed by the requirement that the inward frictional force μN should supply the centripetal force mv^2/r, where N is the normal reaction of the track on the car. If there is no wing on the car, $N = mg$, and we find

$$v^2(\text{no wing}) = \mu r g.$$

If the wing is present we have an extra downforce given by Bernoulli's theorem:

$$\frac{P_A}{\rho} = \frac{P}{\rho} + \frac{1}{2} v^2,$$

with P_A = atmospheric pressure and P the pressure below the wing. Thus $N = mg + A(P_A - P) = mg + \frac{1}{2} A \rho v^2$. With again $mv^2/r = \mu N$ we get

$$v^2(\text{wing}) = \frac{\mu mg}{m/r - \mu A \rho/2}, \tag{1}$$

which of course reduces to the previous formula if the downforce is absent (formally if $A = 0$). With the data given we find $v(\text{no wing}) = 80$ km/h, $v(\text{wing}) = 82$ km/h. Although this is small, it is a significant advantage, so the size and pitch of wings is strictly regulated in motor sport.

S374. Tight corners have small r, while gentle ones have large r. From equation (1) of the last answer we see that on gentle corners the "wing" term $\mu A \rho/2$ is more nearly comparable to the other term m/r in the denominator, and thus has a greater effect. Wings give less advantage on tight corners because the lower speeds make the Bernoulli effect less important.

S375. Using the ideal gas law in the form

$$\frac{P_1 V_1}{T_1} = \frac{P_2 V_2}{T_2}$$

with $V_2 = 2V_1, P_2 = 2P_1$, we get $T_2 = 4T_1$ for the relation of absolute temperatures. With $T_1 = t_1 + 273 = 289$ K we find $T_2 = 1156$ K, or $t_2 = 883°C$.

S376. The ideal gas law can be expressed as

$$P = \frac{R}{\mu}\rho T,$$

where $(R/\mu)T$ is constant at a fixed temperature (μ, the mean molecular weight, is fixed by the gas composition). Thus $P/\rho = $ constant, or $PV/m = $ constant in our case. Hence, writing V_A for 1 liter,

$$\frac{PV}{m_1} = \frac{P_A V_A}{m_2},$$

or

$$P = \frac{m_1}{m_2}\frac{V_A}{V}P_A = \frac{0.858}{0.0015}\frac{1}{12} \times 10^5 = 4.77 \times 10^6 \text{ N m}^{-2}.$$

S377. If the hydrogen pressure is P_H we have

$$P_A + \frac{mg}{S} = P_H,$$

where S is the cross-sectional area of the container, i.e. $S = V_H/h$. Thus

$$P_A = P_H - \frac{mgh}{V_H}.$$

We can find the hydrogen pressure from the equation of state: under the stated conditions hydrogen behaves as an ideal gas, so

$$P_H = n_H \frac{RT}{V_H},$$

where R is the gas constant, n_H the number of moles of molecular hydrogen in $m_H = 0.17$ g and T the temperature. Hence

$$P_A = (n_H RT - mgh)\frac{1}{V_H}.$$

Now $n_H = 0.17/2 = 0.085$ as the molar mass of molecular hydrogen is 2 g. Thus

$$P_A = (0.085 \times 8.31 \times 300 - 21 \times 9.8 \times 0.4)\frac{1}{1400 \times 10^{-6}} = 9.26 \times 10^4 \text{ N m}^{-2}.$$

S378. In the horizontal position the air pressure in l_1 is $P = P_A$ and its volume is $V_1 = l_1 A = 4.2 \times 10^{-5}$ m^3.

In the vertical position with an open end (see Figure 1) the pressure in the lower air column is

$$P_1' = P_A + \rho_{\mathrm{Hg}} g l_2.$$

Using Boyle's law (i.e. the ideal gas law with T fixed)

$$P_1' V_1' = P_A V_1$$

with $V' = Al_1', V = Al_1$, we find

$$l_1' = \frac{P_A l_1}{P_1'} = l_1 \frac{P_A}{P_A + \rho_{\mathrm{Hg}} g l_2} = 0.42 \frac{10^5}{10^5 + 13600 \times 9.8 \times 0.3} = 0.30 \text{ m}.$$

In the vertical position with the corked end we have (see Figure 2)

$$P_1'' = P_3'' + \rho_{\mathrm{Hg}} g l_2. \tag{1}$$

P_1'' and P_3'' can be related to the initial values P_1, P_3 via Boyle's law

$$P_1'' V_1'' = P_A V_1, \quad P_3'' V_3'' = P_A V_3,$$

with $V_1 = Al_1, V_3 = Al_3, V_1'' = Al_1'', V_3'' = Al_3''$. Thus

$$P_1'' = P_A \frac{l_1}{l_1''}, \quad P_3'' = P_A \frac{l_3}{l_3''}.$$

Now $l_3 = l_1$ and $l_3'' = l - l_2 - l_1''$, so from the second relation

$$P_3'' = P_A \frac{l_1}{l - l_2 - l_1''}.$$

Substituting for P_1'', P_3'' in (1) gives

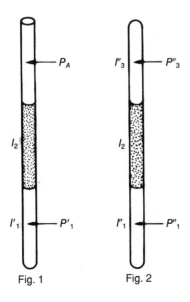

Fig. 1   Fig. 2

$$P_A \frac{l_1}{l_1''} = P_A \frac{l_1}{l - l_2 - l_1''} + \rho_{Hg}gl_2$$

or

$$10^5 \frac{0.42}{l_1''} = 10^5 \frac{0.42}{1.14 - 0.3 - l_1''} + 13{,}600 \times 9.8 \times 0.3,$$

where l_1'' is expressed in meters, i.e.

$$\frac{1}{l_1''} = \frac{1}{0.84 - l_1''} + 0.952$$

or

$$l_1''^2 - 2.94 l_1'' + 0.88 = 0,$$

with the solution $l_1'' = 0.34$ m. (The other root of the quadratic for l_1'' is greater than l, so not admissible as a solution.)

S379. We can regard the air trapped in the bulb and tube as an ideal gas, so that $PV/T = $ constant, or

$$\frac{P_1 V_1}{T_1} = \frac{P_2 V_2}{T_2} \tag{1}$$

where the suffixes 1 and 2 refer to the two states described in the problem.

In state 1 we have

$$P_1 = P_A,$$

since the mercury drop is in equilibrium under the pressure P_1 on one side and atmospheric pressure on the other. Further, V_1 is given by the volume of the bulb and the length l_A of the tube, i.e.

$$V_1 = \frac{4\pi}{3} R^3 + A l_A.$$

Finally, the absolute temperature

$$T_1 = t + 273.$$

In state 2, l_A and t take new values l_A', t', and the trapped air pressure P_2 must increase to balance the extra pressure exerted by the weight of the mercury, i.e.

$$P_2 = P_A + \rho_{Hg} g l_H.$$

From the data given we thus have $V_1 = 1.75 \times 10^{-5}$ m^3, $T_1 = 283$ K,

$$P_2 = P_A + 13{,}600 \times 9.8 \times 0.06 = P_A + 8000 \text{ N m}^{-2},$$

$V_2 = 1.68 \times 10^{-5}$ N m^{-2}, $T_2 = 293$ K. Inserting these in (1) gives

$$\frac{P_A \times 1.75 \times 10^{-5}}{283} = \frac{(P_A + 8000) \times 1.68 \times 10^{-5}}{293}$$

or

$$P_A = 0.93(P_A + 8000),$$

i.e. $P_A = 0.93 \times 8000/0.07 = 1.06 \times 10^5$ N m^{-2}.

S380. The trapped air obeys the ideal gas law at constant temperature (Boyle's law), i.e.

$$PV = P'V' \tag{1}$$

where $P, V; P', V'$ are its pressure and volume in the two states. In the first state, P must balance the pressure at depth h_2 below the surface of the mercury bath, i.e.

$$P = P_A + \rho_{Hg}gh_2,$$

and the volume

$$V = (h_1 + h_2)A,$$

where A is the tube's cross-sectional area. In the second case the mercury level in the tube is higher than that in the bath, so now the pressure in the tube plus that of the mercury column h'_2 balances P_A, i.e.

$$P' = P_A - \rho_{Hg}gh'_2;$$

and the volume

$$V' = (h'_1 - h'_2)A.$$

Substituting in (1) we get

$$(P_A + \rho_{Hg}gh_2)(h_1 + h_2) = (P_A - \rho_{Hg}gh'_2)(h'_1 - h'_2).$$

or, with the data given

$$(P_A + 13{,}600 \times 9.8 \times 0.15) \times 0.2 = (P_A - 13{,}600 \times 9.8 \times 0.15) \times 0.3$$

i.e. $P_A = 5 \times 13{,}600 \times 9.8 \times 0.15 = 10^5$ N m^{-2}.

In the final part we have the state shown in Figure 1 of P380, but with $h_2 = 0$ and $h_1 = h$. The trapped air pressure is now $P'' = P_A$, and its volume $V'' = hA$. From Boyle's law again, we can find h by setting $P''V'' = PV$, where we can use the answer to the first part to evaluate PV. Then

$$P_A h = (P_A + \rho_{Hg}gh_2)(h_1 + h_2).$$

Thus

$$h = \left(1 + \frac{\rho_{Hg}gh_2}{P_A}\right)(h_1 + h_2) = (1 + 13{,}600 \times 9.8 \times 0.15/10^5) \times 0.2 = 0.24 \text{ m}.$$

S381. At the base of the cylinder the pressure given by the trapped air and the column y of mercury must equal that in the mercury bath at depth H, i.e.

$$P_1 + \rho_{\text{Hg}}gy = P_A + \rho_{\text{Hg}}gH. \tag{1}$$

Also, the trapped air obeys Boyle's law (ideal gas law at constant temperature), so

$$P_A V = P_1 V_1$$

where $V = \pi r^2 h$ (the original air volume), and $V_1 = \pi r^2 (h - y)$, i.e.

$$P_1(h - y) = P_A h. \tag{2}$$

Eliminating P_1 between (1) and (2) gives

$$\frac{P_A h}{h - y} + \rho_{\text{Hg}}gy = P_A + \rho_{\text{Hg}}gH.$$

Multiplying through by $(h - y)$ this gives a quadratic equation for y, which after simplification becomes

$$y^2 - \left(\frac{P_A}{\rho_{\text{Hg}}g} + h + H\right)y + Hh = 0,$$

or with the data given

$$y^2 - 2.24y + 0.5 = 0,$$

with the solutions $y = 0.25$, 2.0. Only the first is physical (the other has $y > h$), so $y = 0.25$ m. P_1 follows easily from (2) as

$$P_1 = P_A h/(h - y) = 2P_A = 1.97 \times 10^5 \,\text{N m}^{-2}.$$

The density ρ follows from Archimedes' principle: the buoyancy force $F_B = V_d \rho_{\text{Hg}} g$ must equal the weight $W = V_c \rho g$, where V_d is the displaced fluid volume = (volume of solid cylinder + trapped air) = $\pi R^2 H - \pi r^2 y$, and V_c is the solid cylinder volume = $\pi R^2 H - \pi r^2 h$. Setting $F_B = W$ gives

$$\rho = \rho_{\text{Hg}}\frac{V_d}{V_c} = \rho_{\text{Hg}}\frac{R^2 H - r^2 y}{R^2 H - r^2 h} = \rho_{\text{Hg}}\frac{1 - 0.25(y/H)}{1 - 0.25 \times 0.5}.$$

Using the result of the previous part, this implies $\rho = 13,600 \times (0.94/0.88) = 14,600$ kg m^{-3}.

S382. After the faucet is opened the total number of moles is the same as before, i.e. $n_{\text{tot}} = n + 2n = 3n$. The total volume is $3V$ so the gas density is $3n/3V = n/V$ moles/unit volume. By the ideal gas law the pressure is

$$P = \frac{n}{V}RT \tag{1}$$

where R is the gas constant and T the absolute temperature. This must also be the pressure in each of the containers, so applying the ideal gas law to them in turn gives

$$P = \frac{n_1}{V_1} RT, \ P = \frac{n_2}{V_2} RT,$$

where n_1, n_2 are the numbers of moles in each container. Combining with (1) now gives

$$\frac{n_1}{V_1} = \frac{n_2}{V_2} = \frac{n}{V},$$

or $n_1 = n(V_1/V) = 2n$, $n_2 = n(V_2/V) = n$. (Equivalently we could argue that the constant temperature and pressure throughout requires a constant gas density, with the same result.)

S383. From the ideal gas law we have $PV = nRT$, with P the pressure in atm, V the volume in liters, n the number of moles, and T the absolute temperature in kelvins. The gas constant appropriate here is $R = 0.082$ atm liter $K^{-1} mole^{-1}$. Thus before the faucet is opened,

$$n_1 = \frac{P_1 V_1}{RT}, \ n_2 = \frac{P_2 V_2}{RT},$$

so that

$$n = n_1 + n_2 = \frac{P_1 V_1 + P_2 V_2}{RT} = \frac{10 + 12}{0.082 \times 275} = 0.976 \text{ mole.}$$

After the faucet is opened the pressure is given by the ideal gas law as $P = nRT/V$, with n as above and $V = V_1 + V_2$ the total volume. Thus

$$P = 0.976 \times 0.082 \times 275/8 = 2.75 \text{ atm.}$$

As this is less than P_{crit} there is no leak of gas.

If the temperature is increased to $T' = 400$ K, the new pressure will be

$$P' = \frac{nRT'}{V} = \frac{PT'}{T},$$

or $P' = 2.75 \times 400/275 = 4$ atm $> P_{crit}$, so gas will leak until the pressure is reduced to exactly the critical value. The new number of moles will then be n', given by the ideal gas law as

$$P_{crit} V = n' RT',$$

leading to $n' = P_{crit} V/(RT') = 0.73$ moles.

☐ HEAT AND THERMODYNAMICS

S384. Both the gasoline and the tank will expand to volumes V'_G, V'_T, with

$$V'_G = (1 + \gamma_G \Delta t) V_G = 1.036 V_G$$

$$V'_T = (1 + \gamma_T \Delta t)V_T = 1.0004 V_T.$$

Substituting $V_G = 0.97 V_T$ into the first equation, we find $V'_G = 1.036 \times 0.97 V_T = 1.0049 V_T$. Thus $V'_G > V'_T$ at $40°$ C. The fuel will overflow. Note that this result is independent of the volumes V_T, V_G of the tank and gasoline, and depends only on their *relative* size.

S385. (a) Each side of the plate increases by a factor $(1 + \alpha \Delta T)$, where ΔT is the temperature increase. Since $\alpha \Delta T \ll 1$, the area increases by a factor $(1 + \alpha \Delta T)^2 \simeq 1 + 2\alpha \Delta T$. Thus the coefficient β of surface expansion is approximately twice the linear coefficient ($\beta \simeq 2\alpha = 8 \times 10^{-6}$ $°C^{-1}$). The increase in surface area $S = 100$ cm^2 is therefore

$$\Delta S = \beta S \Delta T = 8 \times 10^{-6} \times 100 \times 100 = 0.08 \text{ cm}^2.$$

(b) From the definition, the amount of heat absorbed is $Q = Cm\Delta T$, where $m = 100$ g is the mass. Thus

$$Q = 0.386 \times 100 \times 100 \text{ J} = 3860 \text{ J}$$

S386. Consider a cube of the solid. If there is a small temperature rise ΔT, its sides increase from a to $a(1 + \alpha \Delta T)$, so its volume increases from $V = a^3$ to $V + \Delta V = a^3(1 + \alpha \Delta T)^3$. Since $\alpha \Delta T \ll 1$, the rhs is approximately $a^3(1 + 3\alpha \Delta T)$. But by definition this is $V(1 + \gamma \Delta T) = a^3(1 + \gamma \Delta T)$, so we must have $\gamma = 3\alpha$.

S387. By Archimedes' principle the steel cube displaces its own mass of mercury, so it floats to a depth d given by $m = a^2 \rho d$, i.e.

$$d = \frac{m}{a^2 \rho} \tag{1}$$

where ρ is the density of mercury. Before heating, a has the value a_0, and after heating this becomes $a = a_0(1 + \alpha_s T)$, where T is the temperature rise. Simultaneously the density of mercury decreases from ρ_0 to $\rho_0(1 + \gamma_m T)^{-1}$ because the same mass of mercury occupies a larger volume. The equilibrium condition (1) becomes

$$d = \frac{m}{a_0^2 \rho_0} \frac{1 + \gamma_m T}{(1 + \alpha_s T)^2} \approx d_0 \frac{1 + \gamma_m T}{1 + 2\alpha_s T}$$

where d_0 was the original depth, since $\alpha_s T \ll 1$. With the data given we find

$$d = d_0 \frac{1 + 1.8 \times 10^{-4} T}{1 + 2.4 \times 10^{-5} T},$$

which is $> d_0$ and increases with T. The level of the mercury bath rises because of the expansion of mercury, and the cube floats slightly more deeply than before.

S388. The minimum power requirement P is given by neglecting the heat lost to the pipes, etc. To heat a mass m_w of water of specific heat C_w through Δt, we need an amount of heat $Q = m_w C_w \Delta t = 10^3 \times 4200 \times 28 = 1.176 \times 10^8$ J every hour ($m_w = 10^3$ kg, $\Delta t = 28°C$). Thus $P = Q/3600 = 32.7$ kW.

S389. The volume of air in the room is $V = 4 \times 5 \times 2.5 = 50$ m^3. The quantity of heat of the air must increase by $Q = VC_A \Delta t = VC_A(t_2 - t_1)$. Thus $Q = 50 \times 1500 \times (20 - 10) = 7.5 \times 10^5$ J. At 75% efficiency this gives the required heat supply from the element as $E = Q/0.75 = 10^6$ J. Since $E = P\Delta\tau$, where $\Delta\tau$ is the time interval, we have $\Delta\tau = E/P = 10^3$ s $= 16.7$ min.

S390. To boil the water the kettle must supply a quantity of heat $Q = mC\Delta t$, where m is the mass of the water, C its specific heat per unit mass, and Δt the temperature difference. Here $m = 1$ kg, $C = 4200\,\mathrm{J\,kg^{-1}\,°C^{-1}}$, and $\Delta t = 100 - 15 = 85°C$. Thus

$$Q = 1 \times 4200 \times 85 = 3.57 \times 10^5 \text{ J.}$$

As the kettle is only 50% efficient, the electrical energy required is $E = Q/0.5 = 7.14 \times 10^5$ J. We must convert this to kWh, using 1 kWh $= 1000\,\mathrm{J\,s^{-1}} \times 3600$ s $= 3.6 \times 10^6$ J. Hence

$$E = 7.14 \times 10^5/3.6 \times 10^6 \approx 0.2 \text{ kWh.}$$

The price is therefore only 2 cents.

S391. Each molecule has kinetic energy $E_k = m_m v^2/2$, where m_m is the mass of a molecule. The total energy is then $E = NE_k$, where N is the total number of molecules. But $NE_k = Nm_m(v^2/2) = m(v^2/2) = 180$ J.

S392. Let t be the temperature of the drink after the ice has melted. Then the heat gain of the ice to the point where it melts is

$$Q_I = m_I C_I \Delta t_I + m_I L_I = 0.04 \times 2310 \times (0 + 1) + 0.04 \times 3.36 \times 10^5$$
$$= 13{,}532 \text{ J.}$$

The heat gain of the coke (negative, i.e. a heat *loss*) is

$$Q_c = m_c C_w \Delta t_c = 0.2 \times 4200 \times (t - 20) = 840t - 16{,}800 \text{ J.}$$

Since we must have $Q_I + Q_c = 0$ (no heat exchange with the surroundings), this gives

$$13{,}532 + 840t - 16{,}800 = 0$$

or $840t = 3268$, i.e. $t = 3.9°C$.

S393. The heat loss from the animal's surface varies as l^2, while the heat gain from food varies as the animal's volume, i.e. l^3. Hence small animals (small l) have relatively larger heat losses than gains, and so tend to cool down. However, mammals have to maintain constant body temperature, so it is preferable to have large l in polar regions.

S394. Conservation of heat energy implies that the heat lost by the metal block is gained by the calorimeter and the water within it, i.e.

$$m_m C_m(t_m - t) = m_c C_m(t - t_c) + m_w C_w(t - t_c),$$

where $t_m = 100°C$ is the metal temperature before immersion in the calorimeter, and $C_w = 4200 \, \text{J kg}^{-1} \, °C^{-1}$ is the specific heat of water. Thus

$$10 C_m(100 - 51) = 0.25 C_m(51 - 10) + 5 \times 4200(51 - 10),$$

or

$$479.75 C_m = 8.61 \times 10^5,$$

so that $C_m = 1795 \, \text{J kg}^{-1} \, °C^{-1}$. This is about 0.43 of the specific heat of water.

S395. If the temperature rise is $\Delta t \, °C$, the block's heat energy increases by $Q = CM\Delta t$. This is all supplied by the kinetic energy $mv^2/2$ of the bullet, so conservation of energy gives $\Delta t = mv^2/2MC = 0.16\,°C$.

S396. Since the calorimeter is insulated, no heat energy is lost, and the heat gained by the calorimeter and contents must balance that lost by the hot water, i.e.

$$(m_c C_{\text{cu}} + m_1 C_w)(t_3 - t_1) + m_2 C_w(t_3 - t_2) = 0$$

Here C_w is the specific heat of water, which is 1 kcal $\text{kg}^{-1} \, °C^{-1}$ by the definition of the kilocalorie. With the data given we find

$$(0.125 C_{\text{cu}} + 0.06)(45 - 24) + 0.09(45 - 63) = 0,$$

giving $C_{\text{cu}} = 0.137 \, \text{kcal kg}^{-1} \, °C^{-1}$.

S397. With C_w the specific heat of water, conservation of heat energy gives

$$(m_1 + m_2)C_w t = m_1 C_w t_1 + m_2 C_w t_2,$$

since no heat is exchanged with the surroundings. Thus

$$t = \frac{m_1 t_1 + m_2 t_2}{m_1 + m_2} = \frac{1 \times 7 + 2 \times 37}{3} = 27°C.$$

The total internal energy change ΔU is zero since both W (the work done) and $\Delta Q = \Delta Q_1 + \Delta Q_2$ (the total heat absorbed) are zero. However, there is a nonzero entropy change $\Delta S = \Delta S_1 + \Delta S_2$ (entropy of mixing), since the heat transfers $\Delta Q_1, \Delta Q_2 = -\Delta Q_1$ are not performed at the same temperatures. Thus using the second law of thermodynamics, $T\Delta S_1 = m_1 C_w \Delta T$, etc., where T is the absolute temperature, leads to $\Delta S_1 = m_1 C_w \ln(T/T_1)$, etc, and hence

$$\Delta S = \Delta S_1 + \Delta S_2 = m_1 C_w \ln\left(\frac{T}{T_1}\right) + m_2 C_w \ln\left(\frac{T}{T_2}\right)$$

$$= 4200 \ln\left(\frac{273 + 27}{273 + 7}\right) + 2 \cdot 4200 \ln\left(\frac{273 + 27}{273 + 37}\right)$$

$$= 289.8 - 275.4 = 14.4 \text{ J K}^{-1}.$$

Note that the initially hotter component of the mixture *loses* entropy ($\Delta S_2 < 0$), but always less than is gained by the cooler component, so that the total entropy change $\Delta S = \Delta S_1 + \Delta S_2$ is positive as required by the second law.

S398. In an ideal gas $PV/T =$ constant, where T is the absolute temperature. Since V is fixed here we have $P/T =$ constant. Thus $P_1/T_1 = P_2/T_2$ or $T_2 = T_1(P_2/P_1) = 3T_1 = 3(273 + t_1) = 819 \text{ K}$. Hence $t_2 = 819 - 273 = 546°\text{C}$. By definition we have $\Delta Q = m_g C_V \Delta t$ or

$$C_V = \frac{\Delta Q}{m_g(t_2 - t_1)} = \frac{1.25 \times 10^5}{0.05(546 - 0)} = 4579 \text{ J kg}^{-1}°\text{C}^{-1}.$$

S399. We have $\Delta S = \Delta Q/T$, where ΔQ is the heat absorbed and T the constant absolute temperature in the process. But $\Delta Q = m_w L_w$, so that

$$\Delta S = \frac{m_w L_w}{T}.$$

We have to convert L_w to J kg^{-1} by multiplying by the mechanical equivalent of heat (4200 J cal^{-1}), so that $L_w = 2.27 \times 10^6 \text{ J kg}^{-1}$. Also $T = 273 + 100 = 373 \text{ K}$, as water boils at $100°\text{C}$ at atmospheric pressure. Thus $\Delta S = 0.25 \times 2.27 \times 10^5/373 = 1520 \text{ J K}^{-1}$.

S400. The gas only does work when the volume changes, i.e. in the phases $1 \rightarrow 2$ and $3 \rightarrow 4$. Thus the work done by the system is

$$\Delta W = W_{1,2} + W_{3,4} = P_1(V_2 - V_1) + P_3(V_1 - V_2) = (P_1 - P_3)(V_2 - V_1)$$

$$= (P_1 - P_1/2)(2V_1 - V_1) = P_1 V_1/2 = 10^5 \text{ J}.$$

Since the process was cyclic ($\Delta U = 0$), the first law of thermodynamics $\Delta Q = \Delta W + \Delta U$ shows that this must also be the absorbed heat ΔQ, i.e. $\Delta Q = 10^5$ J.

S401. The air obeys the perfect gas law $PV/T =$ constant. Before heating we have $P_1 = 760 \text{ mmHg}$, $V_1 = 7 \text{ l}$, $T_1 = 300° \text{ K}$. (As we only compare ratios we can use mixed units of this kind.) After heating we have $P_2 = 770 \text{ mmHg}$ (remember that the mercury level in the inner arm must have been depressed by 5 mm!), $V_2 = 7.005$ l, (including the volume of air in the pipe, which is $V_p = 10 \times 0.5/1000 = 0.005$ l) and we wish to find T_2.

Equating $P_1 V_1/T_1$ and $P_2 V_2/T_2$ we find

$$T_2 = \frac{P_2 V_2}{P_1 V_1} T_1 = 770 \times 7.005/(760 \times 7) \times 300 \text{ K}$$

$$= 304.16 \text{ K} = 31.2\,°\text{C}.$$

S402. We have $V_1 = 4 \times 10^{-4}$ m^3, $T_1 = t_1 + 273 = 288$ K, $T_2 = 273$ K. In an adiabatic process (no heat exchange with the surroundings) the gas obeys the law $T_2 V_2^{\gamma-1} = T_1 V_1^{\gamma-1}$ (this is equivalent to requiring zero entropy change, i.e. $\Delta S = 0$, since $\Delta Q = 0$). Hence we have

$$V_2 = \left(\frac{T_1}{T_2}\right)^{\frac{1}{\gamma-1}} V_1 = \left(\frac{288}{273}\right)^{2.5} V_1 = 1.14 V_1 = 4.57 \times 10^{-4} \text{ m}^3.$$

During the isothermal (constant-temperature) contraction the ideal gas law gives $PV = nRT$. Since the gas is held at temperature T_2 and reaches pressure P_1, its volume V_3 at that point is given by

$$P_1 V_3 = nRT_2.$$

In the initial state we have

$$P_1 V_1 = nRT_1,$$

so dividing this into the previous equation we get

$$V_3 = \frac{T_2}{T_1} V_1 = 0.95 V_1 = 3.8 \times 10^{-4} \text{ m}^3.$$

Thus the gas has a slightly smaller volume than initially.

S403. In an adiabatic process in an ideal gas we have

$$P_1^{1-\gamma} T_1^\gamma = P_2^{1-\gamma} T_2^\gamma$$

so that

$$T_2 = \left(\frac{P_2}{P_1}\right)^{\frac{\gamma-1}{\gamma}} T_1. \tag{1}$$

Using $\gamma = 5/3$ for a monotomic gas we get $T_2 = T_1(0.1)^{2/5} = 160$ K. As the process is adiabatic $\Delta Q = 0$, so by the first law of thermodynamics, we have the work done $\Delta W = -\Delta U$, where ΔU is the increase in internal energy. For a monotomic gas $U = 3nRT/2$, so $\Delta W = -\Delta U = -3nR\Delta T/2$, where $\Delta T = T_2 - T_1$. Thus $\Delta W = 3 \times 5 \times 8.31 \times (400 - 160)/2 = 14,958$ J.

S404. The air in the tire expands and cools adiabatically as it rushes out of the valve. Equation (1) of the previous problem gives a quantitative estimate; with $P_2 = P_1/6, \gamma = 1.4$ (appropriate for air), and $T_1 = 290$ K, we find $T_2 = 174$ K, or $-99\,°$C! Of course there is very little cool air, so the ice soon disappears. A similar effect causes the tiny cloud of water vapor seen on opening coke or champagne bottles.

S405. On the windward side the air rises; here the pressure is lower, so the moisture-laden air has expanded. The expansion is too rapid for much heat to be lost or gained, so it is effectively adiabatic, and the air cools, causing the water vapor to condense and fall as rain or snow. On the other side, the air falls and is adiabatically compressed, so its temperature rises. This gives a warm dry wind. Another example is the *Föhn* north of the Alps.

S406. The derivation of equation (1) is still valid, so

$$\frac{P_1}{P_2} = \frac{9r_2}{r_1}, \tag{1}$$

but equation (2) is no longer valid, as the gas now expands adiabatically not isothermally. We replace (2) using the adiabatic relation $PV^\gamma = $ constant. Since $\gamma = 5/3$ for a monatomic gas and $V \propto r^3$, this requires $Pr^5 = $ constant, so (2) is replaced by

$$\frac{P_1}{P_2} = \frac{r_2^5}{r_1^5}. \tag{2'}$$

Eliminating P_1/P_2 between (1) and (2') now gives $r_2 = \sqrt{3}r_1$ as opposed to $r_2 = 3r_1$ in the isothermal case. The greater expansion in that case results from the fact that energy is being fed into the gas there to keep its temperature constant. This meant that more work could be done expanding the balloon against the tension in the walls.

S407. The change takes place at constant pressure, for which the specific heat is $C_P = C_V + R$ (the extra term R comes from the work done against the pressure). Then

$$\Delta Q = nC_P\Delta T = n(C_V + R)(T_2 - T_1) = \left(\frac{C_V}{R} + 1\right)(nRT_2 - nRT_1).$$

Using the ideal gas law, we can replace nRT_2, nRT_1 by P_0V_2, P_0V_1 respectively, so

$$\Delta Q = \left(\frac{C_V}{R} + 1\right)P_0(V_2 - V_1) = \left(\frac{0.6}{8.31} + 1\right)10^5(0.5 - 1) = -5.36 \times 10^4 \text{ J}.$$

The negative sign shows that heat energy has been lost from the gas.

S408. From the ideal gas law we have $PV_1 = nRT_1$, $PV_2 = nRT_2$, with $n = 2$. Thus $T_1 = PV_1/2R$, $T_2 = PV_2/2R$, and we find $T_1 = 274.3$ K, $T_2 = 640.2$ K. Using the first law of thermodynamics we have

$$Q = \Delta U + \Delta W,$$

where $\Delta U = U_2 - U_1$ is the increase in internal energy, and $\Delta W = P(V_2 - V_1)$, the work done by the gas in the expansion. Since $U = (3/2)nRT = 3RT$ for an ideal monatomic gas, and $PV = nRT = 2RT$, we have

$$Q = 3R(T_2 - T_1) + 2R(T_2 - T_1) = 5R(T_2 - T_1)$$

$$= 5 \times 8.31(640.2 - 274.3) = 1.52 \times 10^4 \text{ J}.$$

The entropy change of an ideal monatomic gas is

$$\Delta S = \frac{3}{2}nR\ln\frac{T_2}{T_1} + nR\ln\frac{V_2}{V_1}$$

so that here $\Delta S = 3 \times 8.31\ln(640.2/274.3) + 2 \times 8.31\ln(0.07/0.03) = 35.2 \text{ J K}^{-1}.$

S409. Heat flows from body 2 to 1 as $T_2 > T_1$. The heat absorbed by body 1 must be exactly that lost by body 2, i.e.

$$0 = \Delta Q_1 + \Delta Q_2 = mC_1\Delta T_1 + mC_2\Delta T_2$$

where $\Delta T_1 = T - T_1$, $\Delta T_2 = T - T_2 = T - 2T_1$. With $C_2 = 1.5C_1$, we get

$$0 = mC_1(T - T_1) + 1.5mC_1(T - 2T_1)$$

i.e. $T = 1.6T_1$.

The entropy changes are

$$\Delta S_1 = mC_1 \ln\frac{T}{T_1},$$

$$\Delta S_2 = mC_2 \ln\frac{T}{T_2} = 1.5mC_1 \ln\frac{T}{2T_1}.$$

Substituting $T = 1.6T_1$, we get $\Delta S_1 = mC_1 \ln 1.6 = 0.47mC_1$, $\Delta S_2 = 1.5mC_1 \ln(1.6/2) = -0.335mC_1$. Clearly $\Delta S = \Delta S_1 + \Delta S_2 > 0$, as required by the second law of thermodynamics. Note that this occurs because in the expression $\Delta S = \Delta Q/T$ it is the body with the smaller value of T which has $\Delta Q > 0$, i.e. heat flows from the hotter body to the cooler body.

S410. The initial volume V_1 is given by using the ideal gas law $P_1V_1 = nRT_1$ (n = number of moles, R = gas constant). For O_2 the molar mass is $m_M = 32$ g, so the number of moles here is $m/m_M = 160/32 = 5$.

Thus

$$V_1 = \frac{nRT_1}{P_1} = 5 \times 8.31 \times 300/10^5 = 0.125 \text{ m}^3.$$

In an adiabatic process we have $P_1 V_1^\gamma = P_2 V_2^\gamma$, or using the ideal gas law, $T_1 V_1^{\gamma-1} = T_2 V_2^{\gamma-1}$, where $\gamma = 7/5$ for a diatomic gas. Thus

$$V_2 = \left(\frac{P_1}{P_2}\right)^{1/\gamma} V_1 = (0.1)^{5/7} \times 0.125 = 0.024 \text{ m}^3$$

and

$$T_2 = \left(\frac{V_1}{V_2}\right)^{\gamma-1} T_1 = (0.125/0.024)^{2/5} \times 300 = 580 \text{ K}.$$

As the process is adiabatic, there is no entropy change, i.e. $\Delta S = 0$. Then using the second law of thermodynamics we have $\Delta U = -\Delta W$, i.e. all of the work done in compressing the gas goes into raising the internal energy of the oxygen. For a diatomic gas we have $U = (5/2)nRT$, so

$$\Delta U = \frac{5}{2}nR\Delta T = 2.5 \times 5 \times 8.31(580 - 300) = 2.91 \times 10^4 \text{ J}.$$

This is also the work done in the compression.

S411. Since the process is isothermal, T does not change, so

$$\Delta T = 0.$$

In an ideal gas at fixed temperature, we have $PV = \text{constant}$, so $PV = P_0 V_0$, or $P = P_0(V_0/V) = P_0/2$ (since $V = 2V_0$). Thus

$$\Delta P = -\frac{P_0}{2}.$$

The internal energy of a fixed mass of an ideal gas depends only on the temperature [$U = (3/2)nRT$, with n the number of moles and R the gas constant]. Thus U does not change, i.e.

$$\Delta U = 0.$$

Using the first law of thermodynamics, we have $\Delta U = Q - \Delta W$, where Q is the heat absorbed by the system and ΔW the work done by it. Here $\Delta U = 0$, so $Q = \Delta W$ and we have $Q = T\Delta S$ (quasistatic process). Thus $\Delta S = \Delta W/T$. Now we use $\Delta W = nRT \ln(V/V_0)$ as given. In our case $V/V_0 = 2$, so

$$\Delta S = nR \ln 2 = 0.693nR.$$

S412. From the first law of thermodynamics

$$\Delta Q = \Delta U + \Delta W,$$

where ΔU is the increase of internal energy of the pump ($\Delta U = 0$ in a cyclic process), $\Delta Q = Q_2 - Q_1$ is the heat absorbed by the "working substance" of the pump, and $\Delta W = -W$ is the work done by it (minus the work done by the motor). Thus

$$Q_2 = Q_1 - W. \tag{1}$$

Now since the process is assumed completely efficient, the entropy change ΔS_2 of the outside "heat reservoir" must be exactly minus the entropy change ΔS_1, i.e.

$$\Delta S_1 + \Delta S_2 = 0.$$

(The entropy of the working substance of the pump does not change as it performs a cyclic process.) Using the second law of thermodynamics we have

$$\frac{Q_1}{T_1} - \frac{Q_2}{T_2} = 0$$

(the outside loses heat, i.e. $\Delta S_2 = -Q_2/T_2$). Thus $Q_2 = Q_1(T_2/T_1)$. Substituting from (1) we get

$$Q\left(1 - \frac{T_2}{T_1}\right) = W$$

so that $Q_1 = (1 - T_2/T_1)^{-1} W = (1 - 268/290)^{-1} W = 13.2W$. (Note that we must convert the temperatures to kelvins, i.e. $T_1 = 17\,°\text{C} = 290$ K.) Thus we get 13.2 J of heat for each joule of electric power used. Even allowing for losses (pump and motor not totally efficient) this method of heating is still much more efficient than normal electrical heating.

S413. The cube has side $l = V^{1/3}$. Consider the impact of a single molecule with a wall perpendicular to the x-direction. If the molecule has x-velocity v_x, an elastic collision with the wall reverses v_x, so the net momentum transferred to the wall is $p = 2mv_x$. Since its x-motion is simply reversed, the molecule will subsequently hit the opposite wall after time l/v_x and return to the first wall after total time $t = 2l/v_x$. Since by Newton's second law, force = rate of change of momentum, the net force on the first wall is $F = p/t = mv_x^2/l$. The total force on the wall is thus

$$F_{\text{tot}} = \frac{m}{l}\Sigma v_x^2.$$

Since the gas molecules have on average the same velocity in all directions, we have $\Sigma v_x^2 = \Sigma v_y^2 = \Sigma v_z^2$, so the rms velocity can be expressed as

$$v^2 = \frac{3}{N}\Sigma v_x^2.$$

Using this and the fact that the wall has area l^2, we find the pressure on the wall is $P = F_{\text{tot}}/l^2$, i.e.

$$P = \frac{Nmv^2}{3V} \tag{1}$$

since $l^3 = V$.

S414. The average speed is

$$v_{\text{ave}} = \frac{1}{3}(1 + 3 + 10) = 4.67 \text{ m s}^{-1}.$$

The rms speed is

$$v_{\text{rms}} = \left[\frac{1}{3}(1^2 + 3^2 + 10^2)\right]^{1/2} = 6.06 \text{ m s}^{-1}.$$

S415. The ideal gas law states that $PV = NkT$, where N is the number density of the gas molecules. Comparing with equation (1) of S413 above, we see that $v^2 = 3kT/m$. Now $m = \mu m_H$, so $v^2 = 3kT/(\mu m_H)$ or

$$v = \left(\frac{3kT}{\mu m_H}\right)^{1/2}. \tag{1}$$

S416. From equation (1) of the previous solution, we have

$$v = [3 \times 1.38 \times 10^{-23} \times 300/(32 \times 1.67 \times 10^{-27})]^{1/2} = 482 \text{ m s}^{-1}$$

for O_2 and four times this value for H_2, i.e. $v = 1928 \text{ m s}^{-1}$, since μ is smaller by a factor of 16.

S417. For typical room dimensions, say 10–20 m, the speeds found in the previous solution imply that the scent is detected virtually instantaneously, i.e. in less than 0.1 s. The fact that the perfume is in reality only slowly detectable shows that the effective speeds of the molecules are far smaller than their rms speeds. This implies that the molecules are constantly deflected from straight-line motion by collisions between themselves. These collisions determine the *transport properties* of the gas (e.g. diffusivity, heat conductivity) for given density and temperature.

S418. The total internal energy of N molecules of monotomic gas is

$$U = N.\frac{1}{2}mv^2,$$

where m is the molecule mass and v the rms speed. Using $v^2 = 3kT/(\mu m_H)$ (see S415), we have

$$U = \frac{3kT}{2\mu m_H}.$$

At constant volume the first law of thermodynamics implies $\Delta Q = \Delta U$, so the specific heat per unit mass at constant volume is

$$C_V = \frac{\Delta Q}{\Delta T} = \frac{\Delta U}{\Delta T} = \frac{3k}{2\mu m_H}.$$

The energy required to heat the same mass of helium and argon through the same temperature is inversely proportional to the mean molecular mass μ. Thus the heat required for the argon sample is $1 \times 4/40 = 0.1$ kJ. Physically this lower value results from the fact that an argon atom is more massive than a helium atom, and so there are fewer argon atoms in the same mass. Since each atom has the same energy $3kT/2$ at a given temperature, less heat is required to raise the temperature of the argon sample.

S419. As the piston moves inwards, molecules hitting it rebound with greater kinetic energies. If the piston moves in at speed u, it sees each molecule elastically reflected at speeds $v_x + u$, so they have x-velocities $v_x + 2u$ in the laboratory reference frame. Collisions between molecules share this extra energy and raise the rms speed v and thus the temperature. If the compression is adiabatic, this happens before any of this extra energy is lost to the surroundings. In summary, the piston does work against the gas pressure, and this heats the gas. The pressure is raised because the momentum transfer between the piston and the gas molecules is increased.

S420. The gas molecules at the base have on average gained kinetic energy mgh compared with those at the top (which have higher potential energy). This raises the pressure at the base by $Nmgh = \rho gh$, where N is the number of molecules per unit volume, i.e. by precisely the amount required to bear the total weight of the gas. Hence the full weight of the gas registers on the scale. The same argument shows that the pressure at every height in the gas is exactly that required to support the weight of the gas above that height.

S421. Equation (1) of S415 shows that $v = (3kT/\mu m_H)^{1/2}$, so the escape temperature T is given by setting this equal to v_{esc}, i.e.

$$T = \frac{\mu m_H v_{esc}^2}{3k}.$$

Thus lighter compounds escape at lower temperatures. With the data given, we find $v = 27.8T^{1/2}$ m s^{-1} for oxygen. For this to reach v_{esc} requires $T = 1.6 \times 10^5$ K. Similarly for nitrogen we find $T = 1.4 \times 10^5$ K, and for hydrogen $T = 1 \times 10^4$ K. This difference is important in explaining why the Earth has lost most of the hydrogen in its original atmosphere, but retains the oxygen and nitrogen.

☐ **LIGHT AND WAVES**

S422. By symmetry we need consider only ray A. It enters the prism normally, and so is undeflected until encountering the glass–air boundary at angle α (see Figure). Let the emergent ray make an angle β to the outward normal. By Snell's law $\sin\beta = n\sin\alpha = 1.414 \times 0.5 = 0.707$, so that $\beta = 45°$. The ray therefore emerges at an angle $\beta - \alpha = 15°$ to its original path. The two rays thus converge at twice this angle, i.e. $30°$.

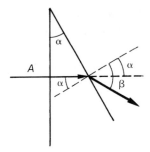

S423. The critical angle for total internal reflection is $\alpha = 90°$ here, so we analyze this case (see Figure). The angle of incidence at AC is the same as angle $\hat{A}$, i.e. $\theta_1 = 60°$. The condition for total internal reflection is (by Snell's law, with the emergent angle $= 90°$) $n\sin\theta_1 = 1$. Thus $n = 1/\sin\theta_1 = 2/\sqrt{3} = 1.1547$. For $\alpha < 90°$, we have $\theta_1 < 60°$ and a part of the light emerges from AC.

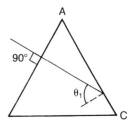

S424. See the Figure. By Snell's law, $\sin 40° = n\sin\alpha_1$. Using $n = 1.3$, we find $\alpha_1 = 29.6°$. Then $x = h\tan\alpha_1 = 0.57$ cm.

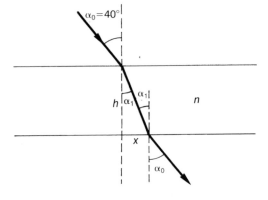

S425. Let the rays from the light source have angle of incidence θ_1 on the water–air interface (see Figure in the problem). Then by Snell's law, $n \sin \theta_1 = \sin \theta_2$. Thus $\sin \theta_1 = \sin \theta_2/n = 0.5/1.3 = 0.3846$, i.e. $\theta_1 = 22.62°$. If the light source is at depth h, we have $\tan \theta_1 = d/h$, or $h = d \cot \theta_1 = 2.4$ m. This compares with the apparent depth $h' = d \cot \theta_2 = 1.7$ m.

S426. By Snell's law we have $\sin \theta_1 = n \sin \theta_2$. The ray is incident on the side of the fiber at angle $\beta = 90° - \theta_2$; for total internal reflection, we require $n \sin \beta \geq 1$, or $n \cos \theta_2 \geq 1$. Thus $n(1 - \sin^2 \theta_2)^{1/2} \geq 1$, or $\sin^2 \theta_2 \leq 1 - 1/n^2$. Hence $\sin^2 \theta_1 = n^2 \sin^2 \theta_2 \leq n^2 - 1$, and the maximum value of θ_1 is therefore $\theta_1^{\max} = \sin^{-1}(n^2 - 1)^{1/2}$. Thus a fiber with $n = \sqrt{2}$ would transmit light incident at any angle, since then $\theta_1^{\max} = 90°$.

S427. The ray path is shown in Figure 1. Clearly $\beta_1 = \beta$, and by Snell's law $\sin \alpha = n(\lambda) \sin \beta$, $n(\lambda) \sin \beta = \sin \alpha'$, so $\alpha = \alpha'$ too. The deflection δ follows from the geometry of Figure 2 as $\delta = 2(\alpha - \beta)$ (exterior angle of triangle ABC). Since $\sin \beta = \sin \alpha/n = 1/(2n)$, we have $\beta(\lambda) = \sin^{-1}[1/2n(\lambda)]$, and so

$$\delta(\lambda) = 60° - 2 \sin^{-1}\left[\frac{1}{2n(\lambda)}\right].$$

Thus for the blue light, $\delta = \delta_b = 60° - 2 \sin^{-1}[1/3.06] = 21.85°$. For the red light, we get $\delta = \delta_r = 60° - 2 \sin^{-1}[1/3.04] = 21.59°$. Hence $\delta_b - \delta_r = 0.26°$. Raindrops can make rainbows!

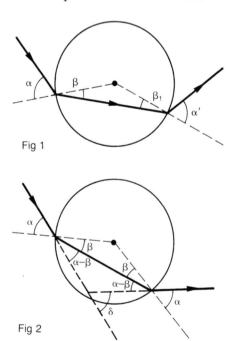

Fig 1

Fig 2

S428. The focal length of a convex mirror is $f = -R/2 = -0.5$ m. The formula $1/s - 1/s' = 1/f$ gives the image distance s' as

$$\frac{1}{s'} = \frac{1}{1.5} + \frac{1}{0.5} = \frac{8}{3}$$

so that $s' = 0.375$ m (i.e. the image is behind the mirror). The magnification is $m = s'/s = 0.375/1.5 = 0.25$, so the image is virtual, upright, and smaller. See Figure for the ray diagram.

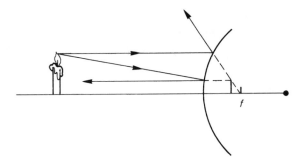

S429. The mirror has focal length $f = -R/2 = 1$ m, so using $1/s - 1/s' = 1/f$ with $s' = -2s$ ($s' < 0$) (since $|m| = |s'/s| = 2$) gives

$$\frac{1}{s} + \frac{1}{2s} = 1.$$

Thus $s = 1.5$ m, $s' = -2s = -3$ m. Since $m = s'/s < 0$, the image is real and inverted. See Figure for the ray diagram.

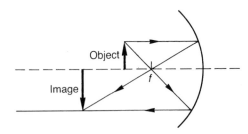

S430. Since $R < 0$, we have $f > 0$ and $f = R/2$. Using the mirror formula with the data given implies

$$\frac{4}{R} - \frac{1}{s'} = \frac{2}{R},$$

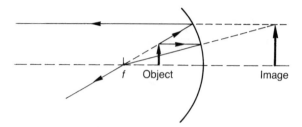

so $s' = R/2 < 0$. The image is a distance $R/2$ behind the mirror and is virtual. Since $m = s'/s = 2$, the image is magnified and upright (see Figure).

S431. The image must be on the same side as the object, i.e. it must be real. Thus for a mirror $s' < 0$, so $s' = -d = -3$ m. Using $1/s - 1/s' = 1/f$ with $s = 0.1$ gives $f = 0.097$ m. Since $R = -2f = -0.194$ m < 0, the mirror must be concave with a radius of curvature of 19.4 cm. From $m = s'/s = -30$ we see that the image is inverted with size $h' = 30 \times 0.5 = 15$ cm. See Figure for the ray diagram.

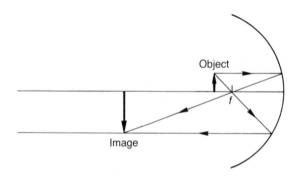

S432. We use the lensmaker's formula

$$\frac{1}{f} = (n - 1)\left(\frac{1}{R_1} + \frac{1}{R_2}\right) = 0.5\left(\frac{1}{R_1} + \frac{1}{R_2}\right).$$

(a) Both radii are positive, so

$$\frac{1}{f} = 0.5\left(\frac{1}{1} + \frac{1}{1.3}\right) = 0.885 \text{ m}^{-1},$$

i.e. $f = 1.13$ m (converging lens).
(b) Both radii are negative, so $f = -1.13$ m (diverging lens).
(c) $R_1 = -1$ m, $R_2 = 1.3$ m, so

$$\frac{1}{f} = 0.5\left(-\frac{1}{1} + \frac{1}{1.3}\right) = -0.115 \text{ m}^{-1},$$

i.e. $f = -8.67$ m (diverging lens).

(d) $R_1 = 1$ m, $R_2 = -1.3$ m, so

$$\frac{1}{f} = 0.5\left(\frac{1}{1} - \frac{1}{1.3}\right) = 0.115 \text{ m}^{-1},$$

i.e. $f = 8.67$ m (converging lens).

(e) $R_1 = \infty$, $R_2 = 1.3$ m. We find $f = 2.6$ m (converging lens).

S433. (a) Using $1/s + 1/s' = 1/f$ with $f = 10$ cm and $s = 5$ cm, we get $s' = -10$ cm. The image is on the same side of the lens (behind the insect) and is virtual. Its size h' follows from $m = -s'/s = 2$, i.e. $h' = 2h$. It is twice the size and upright.

(b) With $f = 10$ cm and $s = 15$ cm, the lens formula now gives $s' = 30$ cm. The image is on the far side of the lens from the insect and is real. From $m = -s'/s = -2$, we have $h' = -2h$, i.e. the image is twice the size and inverted. Note that the image suddenly shifts when the object reaches the focal point. See Figures 1 and 2 for the ray diagrams for cases (a) and (b) respectively.

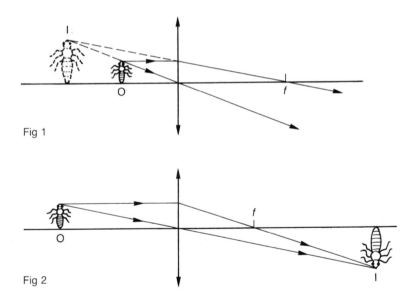

Fig 1

Fig 2

S434. We first find the image created by the lens. Using $1/s + 1/s' = 1/f$ with $f = 0.5$ m and $s = 1$ m, we find $s' = 1$ m. This first image is real and inverted. It forms the object for the mirror, and creates a second image a distance 1 m behind the mirror. This image is virtual, and remains inverted (see Figure 1). This second image itself acts as an object for the lens, at a distance $s = 3$ m. The lens formula gives $s' = 0.6$ m for the resulting image. This third image is

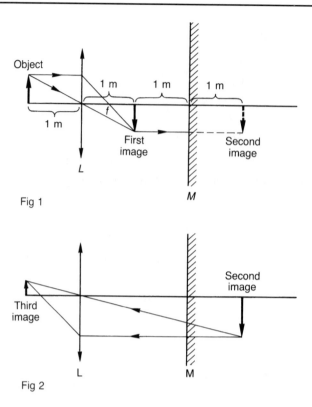

Fig 1

Fig 2

thus on the same side of the lens as the object, real, and upright (see Figure 2). In summary:

First image: real, inverted, 1 m on the opposite side of the lens from the object.

Second image: virtual, inverted, 1 m behind the mirror.

Third image: real, upright, 0.6 m from the lens on the side of the object.

S435. Using the thin lens formula with object distance $s = h - x, s' = x$ gives

$$\frac{1}{h-x} + \frac{1}{x} = \frac{1}{f}.$$

Substituting, we obtain a quadratic equation for x (expressed in cm):

$$x^2 - 50x + 400 = 0.$$

This has the two solutions $x_1 = 40$ cm, $x_2 = 10$ cm. Both of these positions produce a sharp image: exchanging x_1 and x_2 simply exchanges s and s', which must be possible, since they appear symmetrically in the lens formula. The case $x = x_1$ has $s = 10$ cm, $s' = 40$ and has magnification 4, while the opposite case $x = x_2$ has $s = 40$ cm, $s' = 10$ and magnification 0.25.

S436. From the definitions, $p = s - f, p' = s' - f$, so

$$pp' = ss' - (s + s')f + f^2.$$

But multiplying through the thin lens formula by $ss'f$ shows that $ss' = (s + s')f$. Hence the first two terms above cancel, and $pp' = f^2$. This form of the thin lens formula was given by Newton.

S437. We use the fact that the focal length is the image position for an object at infinity (putting $s = \infty$ in the lens formula implies $s' = f$). Thus for the first lens the image is at $s'_1 = f_1$. This forms the object for the second lens, with position $s_2 = -s'_1$ (sign conventions ensure that this expression holds in all cases). Hence $s_2 = -f_1$, so using

$$\frac{1}{s_2} + \frac{1}{s'_2} = \frac{1}{f_2}$$

we find

$$\frac{1}{s'_2} = \frac{1}{f_1} + \frac{1}{f_2}.$$

But s'_2 is the image position for an object at infinity for the combined lens, i.e. its focal length f. Thus

$$\frac{1}{f} = \frac{1}{f_1} + \frac{1}{f_2}.$$

S438. The power is $1/f$, where f is the focal length, and is measured in diopters (meters^{-1}) if f is in meters. By the previous answer, the powers of lenses placed in contact simply add, so the combined lens has power $P = P_1 + P_2 = 2.5$ diopters.

S439. Using the lensmaker's formula

$$P = \frac{1}{f} = (n - 1)\left(\frac{1}{R_1} + \frac{1}{R_2}\right)$$

we get

$$P_A = \frac{2(n_A - 1)}{R}, \quad P_B = -\frac{(n_B - 1)}{R},$$

and so

$$P = P_A + P_B = \frac{1}{R}[2n_A - n_B - 1].$$

With the data given, we see that $P = 0.4/R$ at all three wavelengths. Doublets are often used to correct chromatic aberration, i.e. the variation of focal length with color.

S440. Let the lens–film distance be s'. With $s = \infty$ (distant objects), the lens formula gives $s' = f = 5$ cm if the image is to be in focus.

If the objects are at $s = 1$ m, the lens formula gives

$$\frac{1}{s'} + \frac{1}{100} = \frac{1}{5},$$

i.e. $s' = 5.26$ cm. The lens must be moved 0.26 cm away from the film.

S441. If $s \gg f$ we find from $1/s + 1/s' = 1/f$ that $s' \approx f$ (see previous solution). The magnification is thus $m = -s'/s \approx -f/s$ (the minus sign means that the image is inverted). Hence to change magnification we have to change lenses, so cameras often have interchangeable lenses. For very high magnification, we need very long focal length lenses (which have to be placed further from the film). As the film is the same size, higher magnification lenses have smaller fields of view.

S442. The effective diameter of the lens has been reduced by a factor 2 and therefore the area by a factor 4. The rate at which light illuminates the film is reduced by the same factor, so the photographer must increase the exposure time from 0.02 s to 0.08 s.

S443. We have $s' = 2.5$ cm in all cases (fixed retina–lens distance), while s ranges over $d_n < s < \infty$. Thus $1/s$ has the range $1/d >_n 1/s > 0$. Using the lens formula $1/s = 1/f - 1/s'$ with f and d_n measured in cm, we find

$$\frac{1}{2.5} < \frac{1}{f} < \frac{1}{2.5} + \frac{1}{d_n}.$$

With $d_n = 25$ cm, we get 2.27 cm $< f <$ 2.50 cm. The eye muscles must be able to alter f (and therefore the radius of curvature of the lens) by a factor $2.5/2.27 = 1.1$ (i.e. by 10%).

S444. The person is short-sighted. Her vision can be corrected by placing a lens in front of the eye such that an object at infinity produces an image at a distance $\leq d_f$. Thus for this lens $s = \infty, s' = -1$ m (the image has to be in front of the eye so as to serve as an object for its lens). The lens formula then gives $f = s' = -1$ m. This is a diverging lens, with power $P = -1$ diopters (m^{-1}).

S445. The man is long-sighted. When an object is at $d_n' = 0.25$ m, it must appear to be at $d_n = 0.6$ m, i.e. it must form a virtual image there. Using the thin lens formula with $s = d_n', s' = -d_n$, we find the required focal length f or power P,

$$P = \frac{1}{f} = \frac{1}{d_n'} - \frac{1}{d_n} = 2.33 \text{ diopters.}$$

Thus he needs glasses with converging lenses of focal length $f = 0.43$ m. In most people the near point retreats with age. Reading glasses are required at the latest by the age at which it reaches the length of the arms!

S446. The distance between two objects subtends a larger angle the closer they are to the eye; but the eye cannot focus properly if they are placed closer than the near point. Thus the smallest scale s that the man can distinguish must subtend the minimum angle θ_0 at the near point, i.e. $s = \theta_0 d_n = 0.125$ mm. Note that this formula is correct with θ_0 in radians.

S447. The object is placed just inside the focal point so that it produces a very distant virtual image (see Figure), which can be viewed with comfort. The angular magnification $M = \theta_l / \theta_u$, where θ_l is the angular size of the image as seen through the lens, and θ_u that seen by the unaided eye at the near point. From the Figure, and assuming that $h \ll f, d_n$, we have $\theta_l \approx h/f$, while $\theta_u = h/d_n$. (These results use the facts that $\tan \theta \approx \theta$ for very small angles θ expressed in radians and that the object is very close to the focal point.) Since the power $D = 1/f$, we have $M = d_n/f = d_n D = 2.5$ with the data given.

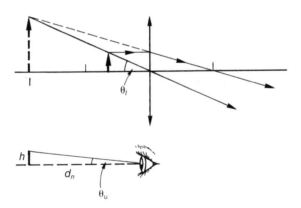

S448. The specimen is very close to the focal point of the objective (see Figure), so the linear magnification of the objective is

$$m = -\frac{s_1'}{s_1} \approx \frac{s_1'}{f_1},$$

where s' is the distance of the real image from the lens.

This magnified real image is the object for the ocular, arranged to be just inside its focal point. The ocular acts as a simple magnifier (see the previous solution), with angular magnification $M_2 = d_n/f_2$, where d_n is the near point of the user's eye. Thus the overall angular magnification is

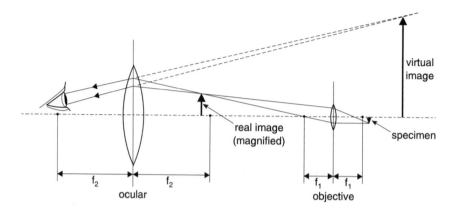

$$M = M_2 m \approx - \frac{s_1' d_n}{f_1 f_2}.$$

Using the thin lens formula $1/s + 1/s' = 1/f$ with $s = s_1 = 0.011$ m, $f = f_1 = 0.01$ m, we find $s_1' = 0.11$ m. Hence with the data given, and assuming $d_n = 0.25$ m, we get $M = -0.11 \times 0.25/0.01 \times 0.05 = -55$. (The minus sign implies that the image is inverted.)

The smallest resolvable detail for a normal unaided eye has a linear size of about $s = 0.125$ mm (see S446 above). With the microscope, this goes down to $s/M = 2.27 \times 10^{-3}$ mm.

S449. The angular magnification of a reflecting telescope is simply $M = -f/f_e$. With the data given, we get $M = 500$. The angular magnification of a refracting telescope is given by the same formula, where f is now the focal length of the objective lens. However, the image for the eyepiece must be formed at the objective's focal point, so a refracting telescope of the same magnification and a similar eyepiece would have to be more than 15 m long.

S450. Comparing with the general formula

$$\psi(x, t) = A \sin \left[2\pi\nu t - \frac{2\pi}{\lambda} x \right]$$

we get $A = 0.1$ m, $\lambda = 5$ m, $\nu = 1/0.01 = 100$ Hz, $v_\phi = \lambda\nu = 5 \times 100 = 500$ m s^{-1}.

S451. Using $v_\phi = \lambda\nu$, we find $\lambda = v_\phi/\nu = 0.5$ m. The wave has the form

$$\psi(x, t) = A \sin \left[2\pi \left(\frac{x}{\lambda} - \nu t \right) \right],$$

so the phase is $\phi = 2\pi(x/\lambda - \nu t)$. At fixed t we have

$$\Delta\phi = \phi_1 - \phi_2 = 2\pi \left(\frac{x_1}{\lambda} - \frac{x_2}{\lambda} \right).$$

Equating this to $\Delta\phi = \pi/6$ rad, we get $x_1 - x_2 = \lambda/12 = 0.0417$ m. Similarly, at fixed x,

$$\Delta\phi = \phi_1 - \phi_2 = 2\pi\nu(t_2 - t_1) = 2\pi\nu\Delta t = 2\pi \times 10^3 \times 10^{-4} = \pi/5 \text{ rad.}$$

S452. The policeman can assume that the car was traveling with a non-relativistic speed v. The corresponding Doppler shift is given by $\Delta\lambda/\lambda_0 = v/c$, where $\Delta\lambda = 6900 - 6000 = 900$ Å and $\lambda_0 = 6900$ Å. Hence the car would have had to travel at $v = (900/6900)c = 0.13c$, which is an absurdly high speed. The speeding ticket is thoroughly justified.

S453. Let the horn emit frequency ν_0. A stationary observer will hear frequency

$$\nu = \nu_0 \frac{v_s}{v_s + v}$$

where v is the velocity of the source *away* from the observer. If the train's speed is v_t, we have $v = -v_t$ as it approaches and $v = v_t$ as it recedes. Hence the observer hears frequency $\nu_0 v_s/(v_s - v_t)$ as it approaches and $\nu_0 v_s/(v_s + v_t)$ as it recedes. The ratio of these two frequencies must be 1.2, so that

$$\frac{v_s + v_t}{v_s - v_t} = 1.2,$$

i.e. $v_t = 0.091 v_s = 30$ m s^{-1} = 108 km/h.

S454. (a) Using $v_s = \nu\lambda$, we find $\lambda = 340/500$ m $= 0.68$ m.

(b) The Doppler effect formula implies

$$\nu = \nu_0 \frac{v_s}{v_s - v}$$

for a source moving towards the pedestrian (the sign of v is reversed for motion away from the pedestrian). Hence

$$\nu = 500 \frac{340}{340 - 40} \text{ Hz} = 566.7 \text{ Hz}$$

and the pedestrian hears the horn at a higher pitch.

S455. The Doppler effect formula relates the wavelength change $\Delta\lambda$ to the source velocity component v along the line of sight – the so-called *radial velocity* by

$$\frac{\Delta\lambda}{\lambda_0} = \frac{v}{c}.$$

(cf. S452 above). Velocity components perpendicular to the line of sight (i.e. in the plane of the sky at that point) have no effect (unless they are comparable with c). Using the formula and the data given, we can conclude that the star has radial velocity

$$v = c \times \frac{6563 - 6562}{6562} = 1.52 \times 10^{-4}c,$$

i.e. $v = 46$ km s^{-1} away from the observer (v is counted positive for motion away, i.e. redshifts). We can say nothing about the transverse motion.

S456. The star's radial velocity (see previous solution) will oscillate back and forth periodically. The mean value gives the radial velocity of the center of mass of the binary system. The amplitude of the radial velocity oscillations and Kepler's laws can be combined to constrain or even measure the masses of the stars of the binary.

S457. The wavelength of the emitted sound is $\lambda = v_s/\nu = 1500/3500 = 0.4286$ m. Local maxima appear where constructive interference occurs, i.e. when the path lengths from A and B to the microphone differ by an integer number of wavelengths. This happens at angles θ to the symmetry line (see Figure) such that

$$d \sin \theta_n = n\lambda,$$

where n is a positive integer. (This formula holds when $d \ll L$ as is the case here.) Since $d = 1$ m, we have $\sin \theta_n = 0.4286n$. We thus have solutions up to $n = 2$, i.e. $\theta_0 = 0$; $\sin \theta_1 = 0.4286$ or $\theta_1 = 25.38°$; $\sin \theta_2 = 0.8572$ or $\theta_2 = 59°$. The detector should thus be placed at distances $x_n = L \tan \theta_n$ from the symmetry line, i.e. at $x_0 = 0, x_1 = 474.4$ m, or $x_2 = 1664$ m.

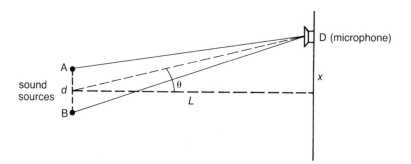

S458. We get destructive interference, i.e zero sound intensity, when the path lengths from A and B differ by exactly half a wavelength, i.e.

$$d \sin \theta_m = \frac{2m - 1}{2}\lambda, \quad m = 1, 2, 3, \dots$$

(the paths differ by an odd number of half-wavelengths). With $d = 1$ m and $\theta_m = 25.38°$ (specifying the position $x = 474.4$ m), we get

$$\frac{2m - 1}{2}\lambda = 0.4286 \text{ m},$$

so using $\lambda = v_s/\nu$ we have

$$\nu = \frac{2m-1}{2}\frac{1500}{0.4286} = (2m-1)\times 1750 \text{ Hz}.$$

For $m = 1$ and $m \geq 3$, ν is respectively below the minimum frequency and above the maximum frequency; for $m = 2$, we have $\nu = 5250$ Hz, which is the required answer.

S459. Without the sheet the phase difference at the central maximum (on the symmetry line) is zero, i.e. $(2\pi/\lambda)d\sin\theta = 0$, where $\theta = 0$. With the sheet in place (see Figure), this is no longer true because of the change of wavelength inside the sheet. Let the angle at which the total phase difference $\Delta\Phi$ is zero be θ (see Figure).

The total phase change has a geometrical contribution

$$\Delta\Phi_g = \frac{2\pi d}{\lambda}\sin\theta,$$

and dispersion contribution (caused by the different refractive index in the sheet)

$$\Delta\Phi_d = 2\pi\frac{t}{\cos\theta}\left(\frac{1}{\lambda} - \frac{1}{\lambda_t}\right)$$

with $\lambda_t = \lambda/n$. Thus

$$\Delta\Phi = \Delta\Phi_g + \Delta\Phi_d = \frac{2\pi}{\lambda}\left[d\sin\theta - \frac{t}{\cos\theta}(n-1)\right].$$

Equating this to zero the central maximum appears at

$$d\sin\theta\cos\theta = t(n-1),$$

or

$$\sin 2\theta = \frac{2t}{d}(n-1).$$

With the data given, we find $\sin 2\theta = 0.17$ or $\theta = 4.9°$.

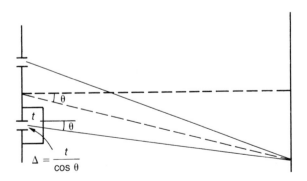

S460. The position of the 8th maximum lies at an angle θ to the center line of the experiment given by $\tan\theta = D/L = 1.46 \times 10^{-2}$. At such small angles $\tan\theta \simeq \sin\theta$ (you may check this using your calculator). Thus $\sin\theta \simeq 1.46 \times 10^{-2}$. At the 8th maximum we have

$$d\sin\theta = 8\lambda,$$

so that $\lambda = d\sin\theta/8 = (3 \times 10^{-4}) \times (1.46 \times 10^{-2})/8 = 5.475 \times 10^{-7}$ m. Using 1 nm $= 10^{-9}$ m, we find $\lambda = 547.5$ nm.

S461. The distance between lines is $d = 1/5000$ cm $= 2 \times 10^{-6}$ m. Maxima appear at angles θ_n satisfying the constructive interference condition $d\sin\theta_n = n\lambda$. For monochromatic light with $\lambda = 6563$ Å $= 6.563 \times 10^{-7}$ m, we get $\sin\theta_n = n\lambda/d = n(6.563 \times 10^{-7}/(2 \times 10^{-6})) = 0.3282n$. Since $\sin\theta_n \leq 1$, the possible values of n are $n = 0, 1, 2, 3$, and the corresponding angles are:

$\sin\theta_0 = 0$, i.e. $\theta_0 = 0$,

$\sin\theta_1 = 0.3282$, i.e. $\theta_1 = 19.16°$,

$\sin\theta_2 = 0.6564$, i.e. $\theta_2 = 41.03°$,

$\sin\theta_3 = 0.9846$, i.e. $\theta_3 = 79.93°$.

For white light we have a range of λ, i.e. $\lambda_1 \leq \lambda \leq \lambda_2$, with $\lambda_1 = 4 \times 10^{-7}$ m, $\lambda_2 = 7 \times 10^{-7}$ m. The second-order interference pattern appears at $\sin\theta_2 = 2\lambda/d$, and thus ranges from

$$\sin\theta_2^{\min} = \frac{2\lambda_1}{d} = 0.4$$

to

$$\sin\theta_2^{\max} = \frac{2\lambda_2}{d} = 0.7,$$

where we have used $d = 2 \times 10^{-6}$ m as before. The third-order pattern appears at $\sin\theta_3 = 3\lambda/d$. Clearly this pattern extends beyond $\sin\theta_2^{\max}$. (Only a part of the wavelength range appears in this order, as $\sin\theta_3$ cannot exceed 1.) The smallest angle for the third order is given by

$$\sin\theta_3^{\min} = \frac{3\lambda_1}{d} = 0.6.$$

Thus the overlap occurs at angles such that $\sin\theta_3^{\min} = 0.6$ and $\sin\theta_2^{\max} = 0.7$, i.e. between $\theta_3^{\min} = 36.87°$ and $\theta_2^{\max} = 44.43°$.

S462. Zeros of the diffraction pattern from a slit appear at angles θ_m satisfying $D\sin\theta_m = m\lambda$ with $m = \pm1, \pm2, \dots$. Thus $\sin\theta_m = (6.870 \times 10^{-5}/10^{-4})m = 0.687m$. Since $\sin\theta_m \leq 1$, zeros are possible only for $m = \pm1$, i.e. for $\theta_{1,-1} = \pm\sin^{-1} 0.687 = \pm43.4°$. If D is doubled the zero condition becomes

$\sin \theta_m = 0.3435m$, so that both $m = \pm 1$ and $m = \pm 2$ are now possible. The first zeros appear at $\theta_{1,-1} = \pm 20.1°$ and the second at $\theta_{2,-2} = \pm 43.4°$, i.e. the positions of the first zeros before D was doubled.

S463. The first zeros of the diffraction pattern are at angles θ such that $D \sin \theta = \pm \lambda$, or $\sin \theta = \pm \lambda/D$, where D is the slit width. Since $\Delta z/2 \ll L$ these angles are small (see Figure) and we may put $\lambda/D = \sin \theta \approx \tan \theta = \Delta z/(2L)$. Thus $D = (\lambda/\Delta z) \times 2L = (7 \times 10^{-5}/1.4) \times 2 \times 100 = 10^{-2}$ cm.

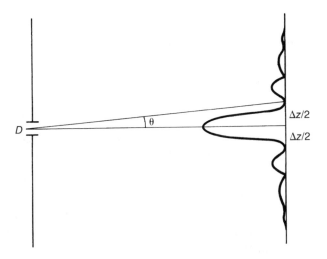

S464. A fraction R_1 of the incident beam is reflected at the air–oil interface. The remaining part of the beam passes through the oil and is reflected at the bottom, producing a second reflected component R_2 (for clarity the two beams are drawn at a slight angle to the vertical in the Figure, although in reality they are both vertical). Minimum (maximum) intensity occurs when R_1, R_2 interfere destructively (constructively). The path phase difference between the two beams is $\Delta\phi = (2\pi/\lambda').2d$, where $\lambda' = \lambda/n$ is the wavelength in the oil, because R_2 travels a distance $2d$ in the oil, unlike R_1. At the upper reflection there is an additional phase shift of π, since $n > 1$, and the reflec-

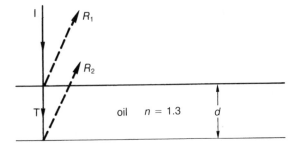

tion at the bottom is total, but this produces no further phase difference. For destructive (constructive) interference, we require $\Delta\phi$ to be an odd (even) multiple of π. Hence for minimum reflected intensity we require

$$\frac{4\pi d}{\lambda/1.25} = (2l - 1)\pi,$$

and for maximum intensity

$$\frac{4\pi d}{\lambda/1.25} = 2l\pi,$$

with $l = 1, 2, 3, \ldots$ in both cases. These lead to

$$d = \frac{2l - 1}{5}\lambda, \quad d = \frac{2l}{5}\lambda$$

respectively.

S465. For minimum visibility we require maximum reflection. From the previous problem this requires a phase difference $\Delta\phi_{\text{req}} = 2l\pi$, with $l = 1, 2, 3, \ldots$ between the rays reflected from the air–paint and paint–glass interfaces. The path difference of the two rays implies a phase difference $\Delta\phi_{\text{path}} = (2\pi/\lambda_p) \times 2d$, which is 14π for the chosen paint thickness. It appears that the engineer will get a promotion. However, at the air–paint interface there is a phase shift of π, since $n_p = 1.4 > 1$, while at the paint–glass interface there is *no* phase shift, since $n_p > n_g$. The true phase difference will thus be $\Delta\phi_{\text{path}} + \pi = 15\pi$, implying destructive interference, minimum reflection, and hence maximum transmission. The official will be very visible, and the engineer may find himself transfered to "other work". Maximal reflection actually requires

$$\Delta\phi_{\text{true}} = \frac{4\pi d}{\lambda_p} + \pi = 2l\pi,$$

or $d = (2l - 1)(\lambda/4)$, i.e. an odd number of *quarter* wavelengths, rather than half-wavelengths as the engineer thought.

S466. Maximum transparency means minimum reflection, which requires

$$\frac{4\pi d_0}{\lambda_s} + \pi = (2l + 1)\pi, \quad l = 1, 2, 3, \ldots$$

where $\lambda_s = \lambda/n$ is the wavelength in the soap film. Thus $d_0 = l\lambda_s/2$, $l = 1, 2, \ldots$, i.e. $d_0 = 5.2 \times 10^{-7} l/(2 \times 1.3) = 2 \times 10^{-7} l$ m, $l = 1, 2, \ldots$. For minimum transparency (maximum reflection) we require instead $d_1 = m(\lambda_s/4)$, with m odd (see previous problem), and also $d_1 < d_0$. For a given l these conditions are first satisfied if $m = 2l - 1$. Thus $d_1 = (2l - 1) \times 10^{-7}$ m for $l = 1, 2, 3, \ldots$.

☐ ATOMIC AND NUCLEAR PHYSICS

S467. We use $\lambda_B = h/p = h/(m_e v_e) = 6.63 \times 10^{-34}/(9.1 \times 10^{-31} \times 10^7) = 7.29 \times 10^{-11}$ m $= 0.729$ Å. (We have used the nonrelativistic expression $p = m_e v_e$ since $(v_e/c)^2 = (10^7/3 \times 10^8)^2 \approx 10^{-3}$, and hence $\gamma = (1 - v_e^2/c^2)^{-1/2} \approx 1$.) Electrons are charged, so an electron beam will be deflected in an electric or magnetic field, while photons are not deflected.

S468. The beam must contain photons with at least enough energy to release an electron, i.e. $E_{\text{photon}} \geq B$, so that $hc/\lambda \geq B$, or $\lambda \leq hc/B = \lambda_{\text{max}} = 6.63 \times 10^{-34} \times 3 \times 10^8/(3 \times 10^{-19}) = 6.63 \times 10^{-7}$ m $= 6630$ Å. This is the maximum possible wavelength for the light beam.

In the case given, the maximum kinetic energy E_{max} follows from Einstein's formula

$$E_{\text{max}} = \frac{hc}{\lambda} - B = \frac{6.63 \times 10^{-34} \times 3 \times 10^8}{4.4 \times 10^{-7}} - 3 \times 10^{-19} = 1.52 \times 10^{-19} \text{ J}.$$

This is equal to 0.95 eV, so the stopping potential is $V_s = 0.95$ V.
These results *do not* depend at all on the light intensity.

S469. Einstein's formula for the photoelectric effect is

$$E_e = \frac{hc}{\lambda} - B,$$

where B is the electron binding energy (work function) of the metal. Since also

$$E_e' = \frac{hc}{\lambda'} - B,$$

we get

$$E_e' - E_e = hc\left(\frac{1}{\lambda'} - \frac{1}{\lambda}\right).$$

Using the values given

$$h = \frac{E_e' - E_e}{c} \frac{\lambda'\lambda}{\lambda - \lambda'} = \frac{(1.35 - 1.02) \times 1.6 \times 10^{-19}}{3 \times 10^8} \frac{5.5 \times 4.8}{5.5 - 4.8} \times 10^{-7},$$

i.e.

$$h = 6.638 \times 10^{-34} \text{ J s}.$$

If the metal has too large a work function for a given light source (i.e. $B > hc/\lambda$), no electrons will be emitted, and the experiment cannot be used to measure h.

S470. At the given frequency each photon has energy $E = h\nu = 6.63 \times 10^{-34} \times 10^6 = 6.63 \times 10^{-28}$ J. The number of photons emitted per second is thus

$$N = \frac{P}{h\nu} = \frac{10^4}{6.63 \times 10^{-28}} = 1.51 \times 10^{31} \text{ s}^{-1},$$

since $10 \text{ kW} = 10^4 \text{ J s}^{-1}$.

S471. The Uncertainty Principle requires $\Delta x \Delta p \geq \hbar$. Thus $\Delta p \geq \hbar/\Delta x$ or $\Delta v \geq \hbar/m_e\Delta x$. Hence the uncertainty in velocity is

$$\Delta v \geq \frac{1.06 \times 10^{-34}}{9.1 \times 10^{-31} \times 10^{-9}} = 1.16 \times 10^5 \text{ m s}^{-1}.$$

S472. The energy is $E = h\nu = 6.63 \times 10^{-34} \times 5 \times 10^{18} = 3.3 \times 10^{-15}$ J $\approx$ 20,600 eV $=$ 20.6 keV. The corresponding momentum is $p = E/c = 3.3 \times 10^{-15}/ (3 \times 10^8) = 1.1 \times 10^{-23}$ N s.

S473. The shortest wavelength λ corresponds to the maximum photon energy, which occurs if an electron hitting the anode gives up all of its energy to radiation. Then

$$\frac{hc}{\lambda} = e\Delta V,$$

i.e. $\lambda = hc/(e\Delta V) = 6.63 \times 10^{-34} \times 3 \times 10^8/(1.6 \times 10^{-19} \times 1.2 \times 10^4) = 1.03 \times 10^{-10}$ m $= 1.04 \text{ Å}$. The number of electrons per second follows from the current, with $t = 1$ s: $N = It/e = 16 \times 10^{-3} \times (1/1.6 \times 10^{-19}) = 10^{17}$ electrons s^{-1}.

S474. To find the Earth's de Broglie wavelength we need its momentum $p_e = M_e v$. Its velocity $v = 2\pi R/t$, with $t = 1$ year $= 365 \times 24 \times 3600 = 3.15 \times 10^7$ s. Thus $p_e = 6 \times 10^{24} \times 2\pi \times 1.5 \times 10^{11}/(3.15 \times 10^7) = 1.8 \times 10^{29}$ kg m s^{-1}. Thus $\lambda_B = h/p_e = 6.63 \times 10^{-34}/(1.8 \times 10^{29}) = 3.7 \times 10^{-63}$ m.

In the Bohr model the angular momentum $l_n = M_e v R$ of the nth orbit is $l_n = n\hbar$. Thus $n = M_e v R/\hbar = (p_e/h).2\pi R = (3.7 \times 10^{-63})^{-1} \times 2\pi \times 1.5 \times 10^{11} = 2.5 \times 10^{74}$. The smallness of λ_B and the enormous value of n show that the classical approximation is excellent.

S475. (a) The energy of an electron is at most $E_{max} = 5 \text{ keV} = 8 \times 10^{-16}$ J. Equating this to the kinetic energy expression $m_e v^2/2$, we find a maximum electron velocity $v = (2E_{max}/m_e)^{1/2} = 4.2 \times 10^7$ m s$^{-1} = 0.14c$, so we can use non-relativistic formulae. Thus

$$\lambda_B = \frac{h}{p} = \frac{h}{m_e v} = 0.017 \text{ nm}$$

(b) In the photoelectric effect the photon energy is at most equal to the electron energy, so $hc/\lambda_{min} = E_{max}$, giving $\lambda_{min} = hc/E_{max} = 0.25 \text{ nm}$.

S476. Using the Compton formula

$$\lambda' = \lambda + \lambda_c(1 - \cos\theta),$$

where $\lambda_c = h/m_e c = 0.024 \text{ Å}$ is the electron's Compton wavelength, and substituting $\cos\theta = \cos 180° = -1$, we get

$$\lambda' = \lambda + 2\lambda_c = 0.2 + 2 \times 0.024 = 0.248 \text{ Å}.$$

Energy conservation requires

$$E_\lambda = E_{\lambda'} + E'_e$$

since the electron's initial kinetic energy ($= E_e - m_e c^2$) is zero. Expressing the photon energies in terms of wavelength ($E = h\nu = hc/\lambda$) we get

$$E'_e = E_\lambda - E_{\lambda'} = \frac{hc}{\lambda} - \frac{hc}{\lambda'}.$$

Thus $E'_e = 6.63 \times 10^{-34} \times 3 \times 10^8[(1/0.2 \times 10^{-10}) - 1/(0.248 \times 10^{-10})] = 1.92 \times 10^{-15} \text{ J} = 1.2 \times 10^4 \text{ eV} = 12 \text{ keV}$.

S477. Let the electron recoil with momentum p at an angle ϕ to the direction of the incident photon, as shown in the Figure. Using the fact that the momentum of a photon is energy/$c = h/\lambda$, conservation of momentum perpendicular and parallel to the incident photon direction in the scattering gives

$$0 = \frac{h}{\lambda_2}\sin\theta - p\sin\phi$$

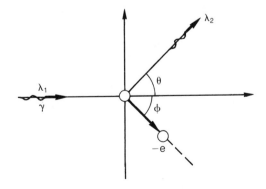

and

$$\frac{h}{\lambda_1} = \frac{h}{\lambda_2}\cos\theta + p\cos\phi.$$

We are told that the electron and photon momenta are equal after the scattering, i.e. $p = h/\lambda_2$. Then the first equation gives $\theta = \phi$. Using this in the second equation and substituting for p yields $\cos\theta = \lambda_2/(2\lambda_1)$.

Now the Compton formula (which uses energy as well as momentum conservation) is

$$\lambda_2 - \lambda_1 = \lambda_c(1 - \cos\theta),$$

with $\lambda_c = h/m_e c$ the electron's Compton wavelength. Substituting for $\cos\theta$ we find

$$\lambda_2 = \frac{\lambda_c + \lambda_1}{1 + \lambda_c/2\lambda_1}$$

Substituting for λ_1 and $\lambda_c = 0.0024$ nm, we find $\lambda_2 = 0.0058$ nm and $\cos\theta = 0.6$, or $\theta = 53°$.

S478. The de Broglie wavelength of an electron in the nth orbit is $\lambda_B = h/p = h/(m_e v_n)$. The ratio of the circumference of the orbit to λ_B is

$$\frac{2\pi r_n}{\lambda_B} = \frac{2\pi r_n}{h}m_e v_n = \frac{m_e v_n r_n}{\hbar} = n,$$

where the last equality follows from the quantization condition.

S479. The electrostatic attraction must supply the centripetal acceleration keeping the electron in a circular orbit of radius r, i.e.

$$\frac{e^2}{4\pi\epsilon_0 r^2} = \frac{m_e v^2}{r},$$

so that

$$r = \frac{e^2}{4\pi\epsilon_0 m_e v^2}.$$

We can eliminate v using the quantization condition in the form $v = n\hbar/m_e r$, to find

$$r = r_n = \frac{4\pi\epsilon_0 \hbar^2}{m_e e^2}n^2$$

The total energy is

$$E_n = \frac{1}{2}m_e v_n^2 - \frac{e^2}{4\pi\epsilon_0 r_n},$$

or substituting for r_n and v_n,

$$E_n = -\frac{m_e e^4}{32\pi^2 \epsilon_0^2 \hbar^2} \frac{1}{n^2}.$$

The coefficient of $1/n^2$ here is the ground state energy E_0 (the Rydberg, $= 13.6$ eV, not to be confused with the Rydberg *constant*).

S480. Balmer lines involve transitions down to level $n = 2$, so the electron must excite a hydrogen atom to level $n = 3$ at least, from the ground state. The required energy is

$$\Delta E = E_0 \left(\frac{1}{1^2} - \frac{1}{3^2} \right) = 13.6 \left(1 - \frac{1}{9} \right) \text{ eV} = 12.09 \text{ eV}$$

($E_0 = 13.6$ eV).

S481. Using the Rydberg formula

$$\Delta E = h\nu = hc/\lambda = E_i - E_f = E_0(n_f^{-2} - n_i^{-2}),$$

where i, f refer to the initial and final states of a Bohr atom, we get

$$\frac{1}{\lambda} = R \left(\frac{1}{n_f^2} - \frac{1}{n_i^2} \right)$$

with $R = E_0/hc = 1.09 \times 10^7$ m^{-1} = Rydberg *constant*. Thus

$$\lambda = \frac{1}{R[(1/1) - (1/16)]} = 9.79 \times 10^{-8} \text{ m} = 979 \text{ Å}.$$

The momentum of the photon is

$$p = \frac{h}{\lambda} = \frac{6.63 \times 10^{-34}}{9.79 \times 10^{-8}} = 6.8 \times 10^{-27} \text{ N s}.$$

By conservation of momentum this is the magnitude of the momentum of the recoiling atom, which has approximately the mass of a proton, $m_p = 1.67 \times 10^{-27}$ kg. Using the non-relativistic formula for this, we find a recoil velocity

$$v = p/m_p = 6.8 \times 10^{-27}/(1.67 \times 10^{-27}) = 4.07 \text{ m s}^{-1}.$$

The fact that $v \ll$ speed of light $c = 3 \times 10^8$ m s^{-1} justifies the use of the non-relativistic formula.

S482. The binding energy of the nth orbit is

$$E_n = -\frac{E_0}{n^2},$$

where $E_0 = 13.6$ eV. Thus

$$E_{100} = -\frac{13.6}{(100)^2} = -1.36 \times 10^{-3} \text{ eV}$$

and

$$E_{1000} = -\frac{13.6}{(1000)^2} = -1.36 \times 10^{-5} \text{ eV}.$$

The binding energy of such orbits is extremely small compared with that of low-lying levels, where it is $\sim$ eV. If m is of order 1 (low-lying level) we have

$$E_m = -\frac{E_0}{m^2}$$

and

$$h\nu_{nm} = E_n - E_m$$

with $|E_n| \ll |E_m|$. Thus

$$h\nu_{nm} \approx |E_m|,$$

almost independently of n. The emitted spectrum is a set of lines extremely close to each other in frequency; if there are many such atoms these lines blend to produce an effective continuum.

S483. At level n, the electron's rotation frequency is (using S479 above)

$$\nu_n = \frac{v_n}{2\pi r_n} = \frac{e^4 m}{2\pi (4\pi\epsilon_0)^2 (n\hbar)^3},$$

or in terms of the Rydberg constant,

$$\nu_n = \frac{2Rc}{n^3}.$$

The spectral line arising from a transition n to $n - 1$ has frequency

$$\nu = \left(\frac{1}{(n-1)^2} - \frac{1}{n^2}\right) Rc = \frac{(2n+1)Rc}{n^4 - 2n^3 + n^2}.$$

For very large n this approaches ν_n, as $n^2, n^3 \ll n^4$ and $1 \ll n$.

S484. Since the energy of a photon is $h\nu = hc/\lambda$, the atoms can in principle absorb photons of any wavelength λ satisfying $hc/\lambda = \Delta E$, i.e.

$$\lambda = \frac{hc}{\Delta E},$$

where ΔE is *any* difference between the levels indicated in the Figure in the problem (in practice the excitation conditions may not populate all of these levels). We have $\Delta E = 3.5, 5.3, 7.0; 1.8, 3.5$ (again); and 1.7 eV. Using the formula above this gives absorption lines at 3552, 2346, 1776, 6906, and 7313 Å.

S485. The ionization energy from the ground state is simply

$$-E_1 = \frac{4 \times 13.6}{1^2} = 54.4 \text{ eV}.$$

The photons in the beam of radiation have energies ranging from

$$E_{\text{low}} = \frac{hc}{\lambda_{\text{low}}} = 51.8 \text{ eV}$$

to

$$E_{\text{low}} = \frac{hc}{\lambda_{\text{high}}} = 24.9 \text{ eV}.$$

Absorptions occur when the photon has the energy $E_n - E_1$ required to lift a ground-state electron into an excited level. Thus absorption lines corresponding to transitions from the ground state will appear for n such that $E_n - E_1 < E_{\text{high}}$. Since $E_1 = -54.4 \text{ eV}$, $E_2 = -54.4/2^2 = -13.6$ eV, $E_3 = -54.4/3^2 = -6.04 \text{ eV}$, $E_4 = -54.4/4^2 = -3.4 \text{ eV}$, $E_5 = -54.4/5^2 = -2.176$ eV, we see that this is possible only for $n = 2, 3, 4$; higher levels are irrelevant since $E_n - E_1 > E_{\text{high}}$ for $n \geq 5$. Since the energies of the absorbed photons are $h\nu_{n,1} = E_n - E_1$, their wavelengths are

$$\lambda_{n,1} = \frac{hc}{E_n - E_1},$$

which gives $\lambda_{2,1} = 305$ Å, $\lambda_{3,1} = 257$ Å, $\lambda_{4,1} = 244$ Å. Absorption lines will appear at these wavelengths.

Absorptions will lift electrons to levels 2, 3, and 4, so radiative transitions $4 \rightarrow 3, 4 \rightarrow 2$ and $3 \rightarrow 2$ will occur *in addition* to the transitions $4 \rightarrow 1, 3 \rightarrow 1$, $2 \rightarrow 1$ directly back to the ground state. These emissions will be isotropic, so viewing the experiment from the side we see six emission lines. When viewing along the beam axis the latter three transitions only partially cancel the absorptions, so we see three emission lines and three absorptions.

S486. The radioactive decay law is $N(t) = N_0 e^{-\lambda t}$, or taking the natural logarithm of each side and rearranging

$$\lambda t = -\ln \frac{N(t)}{N_0}. \tag{1}$$

The half-life is defined as the time at which $N(t)$ has fallen to one-half of its original value. Thus taking $N(t)/N_0 = 1/2$ with $t = t_{1/2}$, we find

$$t_{1/2} = \frac{\ln 2}{\lambda} \tag{2}$$

from (1). To determine λ we use (1) and the data given, i.e. that for $t = 60$ h, we have $N(t)/N_0 = 0.07$. Hence from (1)

$$\lambda = -\frac{1}{60}\ln 0.07 = 0.044 \text{ h}^{-1}.$$

Thus from (2) we find a half-life $t_{1/2} = \ln 2/(0.044) = 15.75$ h.

S487. Beta decay involves the conversion of a neutron ($Z = 0, A = 1$) into a proton ($Z = 1, A = 1$), an electron ($Z = -1$, $A = 0$), and an (anti)neutrino ($Z = 0, A = 0$), i.e.

$$_{26}\text{Fe}^{59} \rightarrow {}_{27}\text{Co}^{59} + e^- + \bar{\nu}$$

The latter two particles are lost from the nucleus, so in beta decay Z increases by 1 and A remains the same.

The second part of the question requires use of the decay law

$$N_2 = N_1 e^{-\lambda t}. \tag{1}$$

As before, we can express the half-life $t_{1/2} = \ln 2/\lambda$. Making λ the subject of the formula (1), we have

$$\lambda = \frac{1}{t}\ln\left(\frac{N_1}{N_2}\right) = \frac{1}{30}\ln(1.6) = 1.57 \times 10^{-2} \text{ days}^{-1}.$$

Thus $t_{1/2} = \ln 2/\lambda = 44.2$ days.

S488. We apply the radioactive decay law to each isotope, i.e.

$$N_{238} = N_0 e^{-\lambda_{238} t}; \quad N_{235} = N_0 e^{-\lambda_{235} t}. \tag{1}$$

Here N_0 is the equal number of each isotope in a given sample at $t = 0$, i.e. the formation of the Earth. The λs are related to the half-lives in the usual way, i.e.

$$\lambda = \frac{\ln 2}{t_{1/2}},$$

which gives

$$\lambda_{238} = \ln 2/(4.5 \times 10^9) = 1.54 \times 10^{-10} \text{ yr}^{-1}.$$

$$\lambda_{235} = \ln 2/(7.1 \times 10^8) = 9.76 \times 10^{-10} \text{ yr}^{-1},$$

Dividing the two equations (1) we find

$$\frac{N_{238}(t)}{N_{235}(t)} = e^{-(\lambda_{238} - \lambda_{235})t},$$

or

$$t = \frac{\ln(N_{238}/N_{235})}{\lambda_{235} - \lambda_{238}}.$$

Thus inserting the current value $99.29/0.71$ for the ratio N_{238}/N_{235}, we find the estimate

$$t = \frac{\ln(99.29/0.71)}{9.76 - 1.54} \times 10^{10} \text{ yr} = 6.0 \times 10^9 \text{ yr}$$

for the age of the Earth.

S489. The radioactive decay law is $N(t) = N_0 e^{-\lambda t}$, where $N(t)$ is the number of nuclei at time t and N_0 that at $t = 0$. The half-life is the time for N to reach $N_0/2$, i.e. $t_{1/2} = \ln 2/\lambda$. The activity is

$$A(t) = -\frac{\Delta N}{\Delta t} = A_0 e^{-\lambda t},$$

where A_0 the known activity for living bone ($t = 0$). The recovered bone has $A(t) = 1.96$ decays $\text{min}^{-1} \text{g}^{-1}$, so this equation gives the age t as

$$t = -\frac{\ln[A(t)/A_0]}{\lambda} = -\frac{\ln(1.96/15.3)}{\lambda} = -\frac{\ln 0.128}{\ln 2} t_{1/2} = 1.65 \times 10^4 \text{ yr}.$$

S490. The energy released in fusion is one half of 23.8 MeV, i.e. 11.9 MeV per deuterium nucleus. The number of atoms in 1 g of deuterium is $N_D = 1/m_D$ where $m_D = 2 \times 1.67 \times 10^{-27} = 3.34 \times 10^{-24}$ g is the mass of a deuterium atom, i.e. $N_D = 3.0 \times 10^{23}$ atoms g^{-1}. Thus the energy yield from fusion of one gram is $E_D = N_D \times 11.9 = 3.6 \times 10^{24}$ MeV g^{-1}. Similarly the fission yield from 1 g of U^{235} is $E_U = N_U \times 200 = (1/m_U) \times 200$ MeV, where $m_U = 235 \times 1.67 \times 10^{-24} = 3.9 \times 10^{-22}$ g is the mass of an atom of U^{235}. This gives $E_U = 5.1 \times 10^{23}$ MeV g^{-1}. The fusion yield from deuterium is considerably higher than the fission yield from U^{238}.

☐ RELATIVITY

S491. The Lorentz transformation relates l and l_0 by $l = l_0(1 - \beta^2)^{1/2}$, with $\beta = v/c$. Thus $\beta = [1 - (l/l_0)^2]^{1/2} = (3/4)^{1/2}$, so $v = 0.87c$. Also $\tau = \gamma \tau_0$ with $\gamma = (1 - \beta^2)^{-1/2}$. Here $\gamma = 2$ so $\tau = 2$ s.

S492. The energy E and momentum p are related by

$$E^2 = p^2 c^2 + m^2 c^4.$$

We need $E = 2mc^2$, and hence $p^2 c^2 = 3m^2 c^4$. Substituting $p = \gamma m v$ gives $\gamma^2 m^2 v^2 = 3m^2 c^2$, so that $\beta^2 = 3/\gamma^2 = 3(1 - \beta^2)$, giving $\beta = \sqrt{3}/2$ or $v =$

0.87c. The momentum will equal mc if $\gamma mv = mc$, giving $\gamma = 1/\beta$. Substituting $\gamma = (1 - \beta^2)^{-1/2}$, we find $2\beta^2 = 1$, giving $v = c/\sqrt{2} = 0.71c$.

S493. In the Earth's rest-frame the particle's lifetime is $\tau = \gamma\tau_0$. For the particle to arrive from the Sun during this interval, we must have $v\tau = v\gamma\tau_0 \geq l$, so the minimum required β satisfies

$$l^2 = v^2\gamma^2\tau_0^2 = \frac{\beta^2 c^2 \tau_0^2}{1 - \beta^2},$$

leading to

$$\beta^2 = \frac{1}{1 + (c\tau_0/l)^2} = 1/1.0001 = 0.9999$$

so that $v = 0.99995c$. Thus if the particle travels sufficiently close to c it can reach the Earth despite its very short lifetime (since $\gamma \simeq 100$).

S494. The signal is sent when S_1's clock reads $\Delta t_1 = 1$ hr. At this point the time Δt_2 which has elapsed in S_2 since synchronization is

$$\Delta t_2 = \gamma\Delta t_1 = \frac{\Delta t_1}{(1 - \beta^2)^{1/2}} = 7.09 \text{ hr.}$$

The corresponding distance traveled by S_1 is $l = 0.99c\Delta t_2$, so the signal takes a time $\Delta t_3 = l/c = 0.99\Delta t_2$ to reach the space station. Hence when the signal is received in S_2 the clock reading is

$$\Delta t = \Delta t_2 + \Delta t_3 = 1.99\Delta t_2 = 14.11 \text{ hr.}$$

S495. The relativistic velocity addition formula is

$$V = \frac{v_s + v_m}{1 + v_s v_m/c^2},$$

while the non-relativistic formula is simply $V' = v_s + v_m$. In the first case

$$V = V_1 = \frac{0.6c + 0.5c}{1 + 0.3} = 0.846c, \quad V' = V_1' = 1.1c;$$

and in the second case

$$V = V_2 = \frac{0.001c + 0.5c}{1 + 0.0005} = 0.5007c, \quad V' = V_2' = 0.5010c.$$

In the first case V_1 is very different from V_1', which also has an impossible value according to relativity. In the second case the relativistic and non-relativistic results are not very different, as $v_s v_m \ll c^2$, so that the two addition formulae give similar results.

S496. Momentum conservation gives

$$\gamma(v)mv = \gamma(V)MV. \tag{1}$$

Energy conservation gives

$$\gamma(v)mc^2 + mc^2 = \gamma(V)Mc^2. \tag{2}$$

Here $\gamma(v) = (1 - v^2/c^2)^{-1/2}$, $\gamma(V) = (1 - V^2/c^2)^{-1/2}$. Dividing (1) by (2) we get

$$V = \frac{\gamma(v)v}{1 + \gamma(v)}.$$

Using $\gamma(v) = (1 - 0.8^2)^{-1/2} = 1.667$, we get $V = 0.5c$. With this result we find $\gamma(V) = 1.155$. Hence from (1)

$$M = \frac{\gamma(v)v}{\gamma(V)V}m = \frac{1.667}{1.155} \times \frac{0.8c}{0.5c}m = 2.31m.$$

S497. The total momentum before the collision is

$$p = \gamma m_e v + \gamma m_e(-v) = 0.$$

A single photon would have nonzero momentum, and so cannot be the sole result of the collision. If exactly two photons are created with momenta p_3, p_4, we must have $0 = p_3 + p_4$, i.e. $p_3 = -p_4$. Thus $|p_3| = |p_4|$, i.e. the momenta are equal and opposite. For a photon we have $E_\gamma = |p|c$, so that the energies are equal. Hence energy conservation gives

$$2\gamma m_e c^2 = 2E_\gamma,$$

or $E = \gamma m_e c^2 = 1.25 m_e c^2 = 1.02 \times 10^{-13}$ J. Also $E_\gamma = hc/\lambda$, so $\lambda = 1.95 \times 10^{-12}$ m $= 1.95 \times 10^{-2}$ Å.

S498. We first find the proton's gamma factor from the relation $E = \gamma m_p c^2$. Since $E = 2000$ MeV $= 2 \times 10^9 \times 1.6 \times 10^{-19} = 3.2 \times 10^{-10}$ J, we find $\gamma = 3.2 \times 10^{-10}/m_p c^2 = 2.13$. Thus $v_p = (1 - 1/\gamma^2)^{1/2}c = 0.88c$. Then using the velocity addition formula

$$v_p' = \frac{v_p - v_s}{1 - v_s v_p/c^2} = 0.59c,$$

and the time for the proton to reach Earth (in the Earth's frame) is $t = l/v_p' = 10^{18}/(0.59 \times 3 \times 10^8) = 5.6 \times 10^9$s $\simeq 178$ yr. In the proton's own frame the time is the proper time t_0, and we have $t = \gamma' t_0$, with $\gamma' = (1 - v_p'^2/c^2)^{-1/2}$ here. Thus $t_0 = t/\gamma = 0.81t \simeq 144$ yr.

S499. In the Earth's frame the trip takes a time $t = l/v = 833$ s. For the aliens it takes the proper time $t_0 = t/\gamma = t(1 - v^2/c^2)^{1/2} = 0.8t = 666$ s.

Alternatively one can calculate the distance in the spaceship's frame. This is $l' = l/\gamma = 0.8l$, and l is the proper Earth–Sun distance. The speed of the Earth in the spaceship's frame is $v = 0.6c$, so the trip takes a time (for the "Earth to arrive at the spaceship") given by $t' = l'/v = 0.8l/v = 0.8t$.

S500. When the alien manages to send her message, time $t = \gamma t_0$ has elapsed on Earth, with $t_0 = 1\,\text{yr}$ the proper time and $\gamma = (1 - 0.998^2)^{-1/2} = 15.82$. This gives $t = 15.82\,\text{yr}$, so that the student is then almost 36 years old. The spaceship has by then traveled a distance $l = 15.82 \times 0.998 = 15.79$ light years from Earth. The radio signal travels at the speed of light and arrives at Earth at time $t' = 15.79\,\text{yr}$ after being sent. The student will thus have aged a total amount $t + t' = 31.61\,\text{yr}$, and will be more than 51 years old.

INDEX

The numbers refer to the problems and solutions. For topics within physics, e.g. *electric potential*, see the Table of Contents.